'Tells a wonderful tale... Impressive to make such a complicated history so concise and accessible to a wide readership.'

Mark Beaumont, author of bestselling
The Man Who Cycled the World

'A wild ride through cycling history... richly researched, intelligent and beautifully written.' **Dr Sheila Hanlon, *CyclingUK***

'Brings to life not only the story and origins of the bicycle, but at the same time teaches us of the unprecedented and amazing changes happening in Britain in the 19th Century.' **Rob Lilwall, author of**
Cycling Home From Siberia

'Extremely well-researched... a must-read for anyone interested in the history and development of the bicycle.' **Anna Hughes, author of**
Eat, Sleep, Cycle **and** *Pedal Power*

'Much has been written about cycling as a sport, as a form of transport and as a way to improve fitness, but far less is known about the important role the bicycle played in social change and reform; William Manners' deeply researched and beautifully written book Revolution redresses that balance.'
Chris Sidwells, author of *A Race for Madmen*

'Fascinating... a heart-warming, often humorous depiction of the development of the bicycle and its role in nurturing human relationships, sporting, social, professional and romantic... touches the lives of cycling enthusiasts through the ages.' **Maria Leijerstam author of**
Cycling to the South Pole

'Provides a fascinating insight into the evolution of cycling culture... and reveals just how much the bicycle has taught us socially and emotionally over the years. Long may that continue.'

WILLIAM MANNERS is a keen cyclist who completed his graduate studies in History at the University of York, specialising in late Victorian cycling. He grew up in the Somerset Levels regularly cycling to school and work. He has written articles about cycling for the *Guardian* and currently lives in Yorkshire where he blogs about the history of cycling: **www.thevictoriancyclist.wordpress.com**

CREVOLUTION

How the Bicycle
Reinvented Modern Britain

WILLIAM MANNERS

DUCKWORTH

This edition first published in the United Kingdom by
Duckworth in 2019

Duckworth, an imprint of Prelude Books ltd
13 Carrington Road, Richmond
TW10 5AA United Kingdom
www.preludebooks.co.uk
For bulk and special sales please contact
info@preludebooks.co.uk

A catalogue record for this book is available from the British Library.

Text design and typesetting by Danny Lyle danjlyle@gmail.com

Printed and bound in Great Britain by Clays

9780715653333

Contents

Introduction:
That Curious Vehicle

Arthur Balfour was not a man known for getting carried away. British prime minister from 1902 to 1905, he is widely credited with the remark, 'Nothing matters very much and most things don't matter at all.' One none-too-impressed commentator summed up his personality as follows:

> An attitude of convinced superiority which insists in the first place on complete detachment from the enthusiasms of the human race, and in the second place on keeping the vulgar world at arm's length.[1]

This may be a little harsh. Certainly, Balfour's long-standing habit of speeding back to his Scottish family estate of Whittinghame at the start of every parliamentary summer recess does not suggest a man who had much time for the 'vulgar world' of politics and London life. Once safely removed to his ancestral retreat, however, he lost no time in indulging in an almost superhuman set of sporting passions.

Regarding his home county as the 'paradise of golfers', Balfour often found the time for two rounds a day during the summer months.

Coiner of the phrase 'lawn tennis', he had a couple of courts built at Whittinghame where he practised regularly. And if not in the mood for either of these, he rode his bicycle around the grounds and the surrounding countryside.

Balfour's cycling energies were not just restricted to the pedalling of his machine. President of the National Cyclists' Union (N.C.U.), he served as head of the institution responsible for managing and overseeing Britain's widely popular cycle racing scene. It was during his premiership in 1899 that the N.C.U. celebrated its twenty-first anniversary by holding a lavish coming-of-age dinner. After hurriedly finishing off some late-night parliamentary business, Balfour took responsibility for presiding over the event.

It is safe to say that he rose to the occasion. Standing in front of an audience sprinkled with earls, lords, MPs, and cyclists from across Britain, Balfour celebrated the bicycle as a machine which had given millions of city dwellers 'a breath of country air, a view of country scenery, a knowledge of the splendours, a knowledge of the magnificence which English scenery presents to us'. With the philosophy of 'nothing mattering very much' nowhere to be seen, Balfour proceeded to proclaim:

> If that be so – and I speak to men who are capable of saying
> of their own experience whether it is so or not – then I say
> there has not been a more civilising invention in the memory
> of the present generation than the invention of the cycle.[2]

This was really saying something. After all, those in the audience had lived through a dizzying period of technological innovation and development. The previous thirty years had seen the invention of, among other things, the electric lightbulb, the telephone, the motor car, the refrigerator and the hand-held camera. Each of these inventions would, in the coming decades, achieve huge usage and transform the lives of millions. And yet, during the closing years of the nineteenth century, none came close to capturing the popular imagination in the same way as the bicycle. As one of Balfour's peers put it:

> The man of the day is the Cyclist. The press, the public, the pulpit, the faculty, all discuss him. They discuss his health, his feet, his shoes, his speed, his cap, his knickers, his handle-bars, his axle, his ball-bearings, his tyres, his rims, and everything that is his, down unto his shirt. He is the man of *Fin de* Cycle – I mean Siècle. He is the King of the Road.[3]

To our more modern minds, such attitudes can be rather hard to fathom. It is easy to view bicycles as unremarkable machines, whose basic design and usage contrasts strongly with most other technologies we use in our everyday lives. Practical and useful? Yes. Inspiring and exciting? For the population at large, it seems unlikely.

There are, of course, many reasons why bicycles represented something very different to men and women from this period. Balfour's speech had been preceded by an extraordinary and never-to-be-repeated phase of machine development, innovation and improvement. Looking back on this remarkable story not only helps us view bicycles through the eyes of contemporaries. It also answers what, for many, I am sure is a more pressing question: what on earth did nineteenth-century bicycles look like?

The birth of the bicycle

As with most accounts of how things have come to be, the story of the bicycle's invention and development contains a fair sprinkling of myths and legends. While a supposed sketch of a machine by Leonardo da Vinci is widely considered a fraud, there is scope to link the invention of the bicycle to an even more unlikely source – the eruption of Mount Tambora in the Dutch East Indies (modern-day Indonesia) in 1815. The huge quantities of sulphur dioxide released into the stratosphere following the eruption are known to have caused a dramatic fall in global temperatures, resulting in 1816 being the 'year without a summer' in the northern hemisphere.

The lowered temperatures caused failed harvests and widespread famine across much of Europe. The crisis was particularly severe in what we would recognise today as Germany, which saw the widespread butchering of horses by people desperate to find alternative food sources. If the story is to be believed, it was this event that led Karl von Drais, an eccentric baron working as a forest inspector in Baden, to search for an alternative human-powered means of transport.

This account overlooks the fact that von Drais had been unsuccessfully working on mechanical vehicles for several years previously; however, even if the famine following Tambora's eruption was not the direct cause of his experiments, it appears to have provided an extra impetus, for by 1817 van Drais had manufactured a working model of his 'Laufmaschine' (running machine), which he patented the following year.[4]

The Laufmaschine (or 'Draisienne' as it quickly became known on the Continent) was certainly not a modern-day bicycle. Lacking any kind of pedal power, it consisted of a wooden frame and two wooden wheels, and was propelled by the rider using his feet to push himself along the ground. Von Drais's machine does, however, represent the first real breakthrough in the search for a practical form of person-powered transportation.

Versions of the Draisienne found their way to England in 1818 through the London carriage-maker Denis Johnson. By increasing the size of the wheels, Johnson made it possible to travel further with each stride, and by tinkering with steering and adding an arm rest, he helped improve their rideability. Johnson's machine acquired a range of nicknames in England, from 'hobby-horse' to the less flattering 'dandy-horse' or 'dandy-charger', referring to the young men of means who rode them. Hobby-horse riding briefly enjoyed widespread popularity in England, France and America, driven in no small part by its sheer novelty value. Some were even confident enough in its utility to predict a promising long-term future. Ackermann's *Repository of Arts* commented:

> It is a most simple, cheap and light machine, and is likely to become useful and generally employed in England, as well as in Germany and France... The swiftness with which a person well-practised can travel is almost beyond belief; 8, 9, even 10 miles may be passed over within the hour, on good and level ground.[5]

But despite such high hopes, the hobby-horse craze proved short-lived. For all Johnson's developments, they remained heavy machines that could weigh upwards of fifty pounds. While this was manageable on flat roads and descents, their bulk became a huge encumbrance when users attempted to push themselves uphill. The absence of any sort of suspension also meant they were highly uncomfortable to ride on rough and bumpy roads.

Moreover, to travel on one of these machines meant exposing yourself to a considerable amount of public ridicule. Those who journeyed into the countryside complained of having to endure 'the censures and remarks of the illiberal and illiterate'.[6] In Britain, the sight of members of the aristocracy pushing themselves along on their 'dandy-horses' proved irresistible to caricaturists and provided inspiration for many cartoons mocking the pretentions of the upper classes. Newspapers were also quick to complain about an issue which has proved to have remarkable staying power. According to one article in the *Lancaster Gazette*, entitled 'More Dandy Horses':

> A few days ago, Mr. Wm. Pollard, a gentleman of great respectability... appeared before Mr. Chambers, the sitting Magistrate at this office, to answer an information laid by Richard Dean [a police constable]... Dean stated, that on the preceding day about 2 o'clock, he saw Mr. Pollard driving and wheeling his horse on the foot-pavement... and therefore he seized the horse under the Paving Act... The Magistrate fined Mr. Pollard in the mitigated penalty of 5s. who, after turning his horse out of the office, mounted and rode off at full speed, leaving a crowd of people much astonished at his expertise in the management of his steed.[7]

The technological shortcomings of hobby-horses, combined with public hostility and mockery, meant they quickly went out of fashion. For the next forty years or so von Drais's invention remained the benchmark that people started from when looking to develop a more user-friendly means of personal transportation. It was only in the mid-1860s that two Frenchmen, Pierre Michaux and Pierre Lallement, produced and marketed a machine that advanced the Draisienne. With the benefit of hindsight, the change they made may appear a rather simple and obvious one: placing pedals and cranks on the front wheel. But this development saw the birth of what we might recognise as bicycles today – machines that the rider powered by the turning of pedals.

Michaux and Lallement's invention instigated another craze, spreading from France to the east coast of America before finding its way back over the Atlantic to England. It was particularly conspicuous in major cities, such as Paris, Boston, New York and London, where riding schools quickly appeared to meet the high demand of pupils wanting to learn how to master their new machines. Streets and public spaces became filled with cyclists; in Paris:

> Velocipedes have become quite a social institution… and velocipeding as necessary an accomplishment as dancing or riding. The veloceman, as he styles himself, is to be seen in all his glory careering at full speed through the shady avenues of the Bois de Boulogne or skimming like some gigantic dragonfly over the level surface of the roads intersecting the Champs-Elysees. The converging avenues of the Arc de l'Etoile are amongst his favourite hunting grounds; there he disports himself continually, to the terror of the aged and the short-winded.[8]

As with hobby-horses, the craze for velocipede riding was in no small part driven by its unconventionality; once this wore off 'velocipedomania' quickly came to an end. Although machines built in the style of Michaux and Lallement were undoubtedly more user-friendly than Draisiennes, they still came with their fair share of issues. At seventy pounds they weighed even more than hobby-horses, and their continued lack of suspension meant they could still be excruciatingly

1 French champion velocipede racer Edmond Moret photographed with
his velocipede (c.1869).

uncomfortable to ride – it was with good reason that they came to be widely nicknamed 'boneshakers'.

This is not to understate the impact achieved by early velocipedes. They could be used to cover far greater distances than had been possible on a hobby-horse, and were used by a far greater number of people spanning a wider range of social backgrounds. Many of the early riders in England appear to have been engineers, mechanics, tradesmen and shopkeepers.[9] Velocipede racing also proved a popular spectator sport, with English and French riders crossing the channel to compete in annual meets which drew large audiences.

Furthermore, velocipedes provided the stimulus for a truly remarkable period of bicycle design and development. All-metal wheels with tensioned spokes, developed in 1870, were lighter and provided far better suspension than their wooden counterparts.[10] Manufacturers also began to increase the size of the front wheel and reduce that of the rear, ensuring that a rider moved further forward with each turn of the pedals. This lowered the amount of exertion required to power a machine and did much to increase their speed. It was for these eminently sensible reasons that machines began to be developed which, at first glance, appear anything but sensible: high-wheeled bicycles which later came to be known as 'penny farthings'.

That high-wheeled bicycles represented a significant improvement on all that had preceded them was made apparent in 1870, when James Moore brought the bicycle he had been racing in Paris earlier in the year to England. In his first competitive outing, Moore easily won his two heats, only to be denied victory in the final when he became the first of many people in Britain to suffer a spectacular fall from his machine. Enthused by initial success, however, English firms realised the potential for high-wheelers and quickly began producing them in large numbers. The most successful of these was James Starley and

William Hillman's 'Ariel'. Patented in 1870, it was the first all-metal English bicycle to be mass-produced and weighed around fifty pounds. Costing £8, it was expensive but not exorbitant, and was proudly declared to be 'the lightest, strongest, safest, swiftest, easiest, cheapest, best finished and most elegant of modern Velocipedes'.[11]

The outbreak of the Franco-Prussian War did much to limit the growth of the French cycle industry, and meant that for much of the 1870s English manufacturers were at the forefront of bicycle design and development. Stimulated by a popular racing scene in which innovation and improvement were essential, front wheels slowly became larger until they could exceed sixty inches in diameter. Bicycles themselves became lighter, so that by the end of the 1870s a high-spec racing machine might weigh as little as thirty pounds. Writing in 1877, Charles Spencer could look back and comment on

> ... the enormous difference between riding the early machines and those of the present day... the contrast between them is really so marked that it is visible to the most ordinary observer. It is difficult to realize the fact that the low clumsy affairs first introduced have developed into the light and graceful machines of the present day.[12]

As the most popular type of bicycle on the market, high-wheeled machines became known as 'ordinaries' – it was not until the early 1890s that they acquired the more disparaging nickname 'penny farthings'. They were predominantly used by men in their late teens and twenties from affluent backgrounds who, as well as having the means to purchase an ordinary, proved most prepared to face the very real risks to life and limb that accompanied riding a high-wheeler. The position of a penny farthing rider, on a saddle which was just above the top wheel, ensured they could fully extend their legs while riding and thus deliver maximum force to the pedals. It also meant, however, that their weight was placed dangerously far forward, particularly when heading downhill. Consequently, head-first descents over the handlebars, known as 'headers', were a common occurrence. Looking back in 1896, one writer recalled:

> Even the most expert rider now and again yielded to the force of circumstances, and took 'a header,' in process of which he left his bicycle for a moment with its little wheel reared aloft, reached out his hands to Mother Earth, and kissed her frantically, while his high-tempered steed lay down docile at his side. If unhurt, he would rise with an absent-minded smile, remount and ride on. But so common, and frequently so serious, did this form of accident become, that every bicycler of necessity counted the chances of it as one of the prices he had to pay for the pleasurable excitement of bicycling.[13]

Journeying out into the countryside also came with its own distinct set of perils. In remote rural areas the unexpected appearance of high-wheeled machines was treated with a mixture of wonder, excitement and a quite staggering amount of hostility. Riders frequently reported sticks being pushed through their wheels, clumps of mud being thrown at them and physical assaults from groups of 'roughs'. Court cases highlight how drivers of horse-drawn vehicles also showed an open aggression to those on ordinaries, driving them off the road and even attacking them with the whips they used on their horses. It was not until 1878 that bicycles were given legal status as carriages, which meant that they were protected for the first time by the same rights as other highway vehicles.

However, despite all the dangers that came with it, ordinary riding proved remarkably popular. Indeed, for those who took up the sport, its accompanying risks and perils appear to have been a part of its charm. Writing to *Cycling*, Britain's most read cycle-based magazine during the 1890s, one long-standing devotee looked back and recalled:

> Thirteen years ago, when I first conceived the idea of taking refuge from *ennui* atop of a high wheel, there was some appearance of justification for a young man's aspiration to master the stately, tall machines then in vogue, as to do so certainly required one to face violence, and risk sudden death – two considerations dear to the typical Britain, and essential conditions to the favourable reception by him of any new form of athletic sport.[14]

Drawing in such enthusiasts, cycling clubs proliferated during the 1870s and 1880s, offering sociability as well as mutual protection during weekend excursions. By 1882, 184 clubs were listed as being active in Greater London and nearly 350 more elsewhere in England, with an average size of around thirty to forty members. 1878 also saw the founding of the Bicycle Touring Club, which changed its name to the more recognisable Cyclists' Touring Club (C.T.C.) in 1883. Providing an official voice for all those who participated in the pastime and regularly campaigning across a range of cycle-related issues, the C.T.C. grew phenomenally in the years following its foundation, boasting over 22,000 members by 1886.[15]

This growth was augmented by another development in cycle design. After spending much of the 1870s focusing almost solely on how to improve ordinaries, manufacturers began to turn their attention towards tricycles: three- or even four-wheeled machines which placed the rider much closer to the ground. Most tricycles were propelled by the rider turning the cranks below their feet, which drove one or two chains that in turn powered the larger wheels. Levers at the side of the machine or handlebars at the front provided a means of steering.

Although heavier and slower than ordinaries, tricycles did possess several distinct advantages over their high-wheeled brethren. Their design ensured that they were far easier to mount, as well as being much safer and more comfortable to ride. Combined with the fact that they could easily be used to carry baggage and spare clothes, tricycles were ideal machines for touring and taking longer holidays away from home. Furthermore, in a society highly preoccupied with social status, they did far less to call into question the 'respectability' of the person riding them. While in Ireland it was not uncommon for older men of high social standing – such as doctors, barristers, solicitors and cler-gymen – to ride ordinaries, in England they were far more strongly associated with their younger kinsmen who had not yet made their way in the world.[16] That it could be viewed as improper for older men to join the ranks of high-wheeled cyclists is evidenced by an 1879 article in *The Cyclist*, which advised:

> When men pass a certain age, say five-and-thirty, they have generally made such a position in life that they cannot, without endangering their dignity, join a club comprised of young clerks and other men of business, however good the social position of their parent may be. If a man has grown old in a club, the case is somewhat different, but even then he had better retire in favour of younger men.[17]

By contrast, tricycle riding was a pastime associated with those who occupied a higher 'position in life', including members of the upper middle classes, aristocracy and even royalty. Tricycle riding was given a huge boost when Queen Victoria saw a rider while on holiday in the Isle of Wight in 1881 and subsequently ordered a pair of machines. While it is doubtful that she ever tested her new purchases, the ranks of tricyclists did contain female members. Unlike ordinary riding, tricycling was commonly viewed as an acceptably 'feminine' activity. The fact that it did not require women to straddle a saddle, could be undertaken in skirts and therefore kept their legs hidden, and often took place on 'sociables' where they were joined by husbands and male relatives all ensured the sight of a female tricyclist was, within the cycling community at least, unlikely to stir controversy.

While it can be easy for us to view machines from this period with incredulity, by the mid-1880s cycling had unquestionably established itself as a pastime. Ordinaries and tricycles ensured that it was an activity that could cater for a range of ages as well as both sexes. Cyclists were now better received in rural areas in Britain, with C.T.C.-approved hotels and restaurants catering for travellers at fixed rates. Cycle racing had also established itself as a popular spectator sport capable of drawing thousands of spectators, with the results of races being reported in both the mainstream and the ever-expanding cycling press.[18]

But for all the developments that had occurred in cycle design, inherent problems persisted. While tricycles had opened cycling up to older men as well as women, they were much more expensive than high-wheelers and therefore unaffordable for most. Despite improvements, the basic design of ordinaries meant they remained impractical machines which struggled to make any headway with groups other than testosterone-fuelled young men. The fact that it was nearly impossible to mount or get off ordinaries easily, and the bulk and size of tricycles, meant neither could claim to be an everyday machine suited to activities other than weekend trips and excursions. There was clearly a gap in the market for a new type of machine capable of meeting some of these shortcomings. The question was who, if anyone, could produce it?

The coming of the safety bicycle...

From the early 1880s onwards manufacturers vied to solve a significant engineering challenge in their pursuit of safer, faster machines. Rather than attaching pedals to the front wheel, could they use a chain to turn the rear wheel? The benefits of this system, which could be studied in various forms of tricycle, would not just be in the area of safety. Because the handlebars would no longer be attached to the pedalled wheel, steering would become much easier and require less effort. A chain system would also ensure that power was provided to the driving wheel continuously, rather than just at the start of each rotation when a cyclist pushed down on the pedals.

Despite these potential benefits, up to 1885 early designs met with limited success. This all changed, however, when John Kemp Starley (whose uncle James Starley had been heavily involved in developing the design of both ordinaries and tricycles), along with his partner William Sutton, produced the second version of their 'Rover' safety bicycle. For the first time a machine whose basic design we would recognise today was on the market, with a diamond frame, two equal-sized wheels and a chain-driven power system. As racers riding Starley's Rover began to break records previously held by ordinary riders, its superiority began to be made apparent, ensuring other makers quickly took on and developed its design. This was greatly aided by Starley himself, who as well as sending samples of his safety all over the world, refused to take a patent out on it. Interviewed in 1897, he recalled:

The machine was… an adaptation of various parts of machines already in existence. It was their combination, of course, that was my doing. Other makers soon saw the value of the new cycle and at once proceeded to copy it.[19]

…and pneumatic tyre

By far the most significant advance to the Rover, however, came not from a cycle manufacturer or indeed anyone connected to the bicycle trade. Away from his work as a veterinary surgeon John Boyd Dunlop, Scottish by birth but living and practising in Belfast, was interested in finding ways he could improve his son's tricycle so it would be smoother and easier for him to ride. Dunlop's experiments culminated in him fitting a crude inner tube on the inside of its wheels in 1888, representing the birth of the pneumatic tyre. Dunlop quickly patented and began producing his invention, all the while testing and improving its design.

While many were sceptical about what they saw as wheels fitted with 'German sausages round the rims', it soon became apparent that Dunlop's invention represented a huge advance on the solid or cushioned rubber tyres that had preceded it. Pneumatic tyres did far more to absorb vibrations and improve suspension, which quickly made cycling a much more comfortable and enjoyable experience. By reducing the resistance between a tyre and the road, they also increased the speed a cyclist could travel by about a third. While punctures now became an unwanted nuisance, the invention of the detachable tyre in 1891 at least allowed cyclists to fix these themselves rather than seeking out specialist assistance. Writing on the pneumatic tyre in 1896, the racing cyclist F.T. Bidlake opined that 'no other invention has so popularised

2 Anfield Bicycle Club member Norman Cook posing doing a 'header' with his ordinary (c.1890).

3 Out with the old and in with the new. Norman Cook now looking much happier atop a safety (c.1890).

cycling, for it reduced the labour of propulsion enormously, and minimised the jolting of the wheels to an extraordinary extent'.[20]

The fitting of safety bicycles with pneumatic tyres represents the birth of the 'modern' bicycle. This is not to say that machines ridden during the 1890s were identical to those today. Their steel frames meant they were far heavier, and it was not until the turn of the century that the invention of the derailleur saw bicycles being produced with multiple gears. Late nineteenth-century machines were fixed speed and required their owner to choose a gear size appropriate to themselves upon purchase. By modern standards brakes were also highly inefficient, meaning many chose brakeless machines which they slowed down by back-pedalling or placing a foot on the front tyre.

Nevertheless, in nearly all respects pneumatic-tyred safeties represented a huge advance on previous models and soon became the dominant make on the market. They symbolised a major advance not just in cycle design, but in the much broader history of human travel and transportation. It is important to remember that this was a time when horse ownership was reserved to a privileged few, and early versions of the motor car were used by an even narrower social bracket. After decades of development, the safety bicycle and pneumatic tyre could deliver on the bicycle's original promise: a practical form of personalised long-distance transportation that was within the reach of many.

While at the beginning of the 1890s the most cheaply available safeties cost around £10 (far greater than most people's weekly wage) they had become increasingly affordable by the century's end. Hire purchase schemes meant they did not need to be purchased all in one go, with the cost instead split over several months. Growing competition between manufacturers also saw prices fall, while second-hand safeties could be bought for as little as £2 from an ever-expanding second-hand market. It has been estimated that during the middle years of the 1890s there were around 1.5 million cyclists in Britain – a number which may well have been considerably higher.[21] Writing in 1899, the prominent cycle journalist George Lacy Hillier (a character whom we shall encounter again later) could confidently pronounce:

> What a wonderful thing is the modern cycle! In years to come, when the historian writes of the Victorian age, he will, without doubt, feel himself constrained by the force of circumstances to write at length of the genesis and development of 'the bicycle' – that curious vehicle which in the nineteenth century added new and altogether unequalled powers of locomotion to those already possessed by man, powers which were dependent on man's muscles alone, and which enabled him to travel farther and faster than he has before been able to progress by their use.[22]

There are many ways in which one might take on Hillier's mantle and write the history of these developments. Aside from the fact that the story of the bicycle's development is of course far more intricate and complex than the condensed version above, the social impacts of people acquiring 'new and altogether unequalled powers of locomotion' were incredibly wide-ranging. Take just one area – work.

Cycling significantly expanded the reach of those working in rural areas: as well as being used by nurses and clergymen (the Bishop of Bath and Wells was known to visit parishioners on his machine), bicycles also enabled the Post Office to start making daily deliveries to every household in Britain during the final years of the century.[23] Cycle ownership could also dramatically reduce commuting times. With older models available for knock-down prices during the 1890s, this was something which became increasingly prevalent among working-class men. A piece in *Nairn's News of the World* from 1896 observed:

> To those who may happen to have been about in the suburbs of London in the early morning, the stream of workmen seen passing on machines which have been rendered obsolete from a fashionable point of view by modern improvements is very considerable.[24]

Through their weekend leisure excursions, cyclists also played a crucial role in improving the quality of country roads and turnpikes. The coming of the railways in the 1840s signalled the death knell for stagecoaches and mail coaches, and saw a dramatic fall in rural road traffic. Unsurfaced and with little demand for their use, thoroughfares that had been a source of national pride soon slipped into a state of disrepair. As the first body of travellers to venture back out onto Britain's highways and byways, cyclists quickly became a powerful lobbying group for their improved maintenance and condition. In 1885 the C.T.C. created the Roads Improvement Association which oversaw considerable improvements in rural roads, benefiting horse-drawn vehicles as well as cyclists, not to mention future generations of motorists.[25]

But the impact on road quality and the everyday practicality afforded by the bicycle fail to explain the feverish excitement it generated. Any history of the bicycle must find a way to capture the sheer pleasure and enjoyment which people found when out riding their machines. There are countless numbers of personal accounts that reveal just how much delight the freedoms created by cycling brought to men and women's day-to-day lives. As Elizabeth Priestley, a keen cycle tourist who produced wonderfully lyrical accounts of her excursions, put it:

To glide along at one's sweet will; to feel the delight in rapid motion that is the result of our consciously exerted strength; to skim like a low-flying bird through the panorama of an ever-varying landscape; to know a new-born spirit of independence... to return from a country spin with a healthy appetite, a clearer brain, and an altogether happier sense of life – an altogether unaccountable freshness of spirit; this is to experience something of the joys of cycling.[26]

The joyous freedom and escape made possible by cycling was central to why individuals such as Balfour placed such a high price on it. After decades of rapid industrialisation, which had led to over half of Britain's population living in towns and cities, the bicycle was unique in how it enabled people to travel swiftly away from polluted urban environments and experience 'the magnificence which English scenery presents to us'. Writing in 1892, the 'Grand Old Man' of British politics, William Gladstone, who was soon to begin his fourth spell as prime minister, offered a similarly positive view of the pastime when he commented:

I have noticed with real and unfeigned pleasure the rapid growth of cycling in this country, for not only does it afford to many to whom it would otherwise be unobtainable a healthy and pleasurable form of exercise, but it also enables them to derive all those advantages of travel which, previous to the advent of cycling, were out of their reach... I consider that physically, morally, and socially, the benefits that cycling confers on the men of the present day are almost unbounded.[27]

The extra mobility created by cycling was not simply a source of satisfaction in and of itself. As we shall see in each of the following chapters, riding a bicycle also created a range of new ways for people to indulge in things that they have always loved to do: to dress up and show off, to socialise, to compete, and to flirt and fall in love. The essential story told by this book is how the simple act of pedalling a bicycle unlocked excitement, adventure and the imaginations of the millions of people who came to embrace it.

Admittedly, this is an assertion which requires several caveats. In the period it covers (the early 1870s to the turn of the twentieth century), the term 'people' cannot be said to refer to the population at large, but a rather small portion of it. During this time cycling was largely enjoyed as a middle- and lower-middle-class leisure activity, whose participants stretched from wealthy businessmen and professionals to clerks, shopkeepers and sales assistants. As well as possessing greater disposable income than members of the working class, from the mid-nineteenth century onwards white-collar workers increasingly enjoyed access to Saturday half-holidays as well as paid time off each year. Combined with the fact that they spent most of their weeks leading sedentary lives, it was the middle classes who had 'the time, money, and desire to take cycling up seriously and in large numbers'.[28]

The lively discussions that surrounded cycling were shaped by the values and standards of this section of society. The middle classes had, by the late nineteenth century, come to accept recreation as an important counterbalance to the stresses of work and duty, and many articles on cycling echoed the celebratory sentiments of Balfour and Gladstone. At a time when middle-class identity was centred around the values of hard work and self-improvement, however, the 'unfettered liberty' people now had to pedal away to enjoy themselves in their free time often prompted a more varied and critical response.

Indeed, the ways in which cycling conflicted with middle-class values in this period provide as much of a backdrop to the story as developments in bicycle design.[29]

Pushing the limits of civilisation

While Gladstone was surely right to comment on the 'physical, social and moral' benefits that could be achieved through cycling, such an assertion did overlook the many other uses individuals found for their bicycles. Not surprisingly, a machine which quickly got the blood flowing, and was an excellent means of finding release after busy working weeks, served other ends than physical and moral improvement. One disgruntled newspaper columnist who had taken a weekend trip to the seaside town of Cromer bemoaned how his Saturday night's sleep had been ruined by a 'party of cyclists' who

> … came into the town shortly after midnight, and, from causes arising either from penury or alcohol, decided that the proper place to pass the night was in the public street, and unfortunately they selected the spot directly under my window as the *locale*. For six mortal hours I had to stand it. There were bugle calls, cat calls, blasphemous calls, curse calls, and every other calling except that adopted by a respectable athlete.[30]

Clearly none too impressed by their stamina, the writer reported how the next morning the group were still causing scandal, as they 'indulged in a bathe… under conditions that would be impossible in a well-regulated watering-place'. In addition to being a disruptive influence at the places they stayed, groups of cyclists also caused regular complaints by their conduct when out riding. 'It is fast becoming impossible to take one's exercise on any of the high roads leading out of London... on account of the bands of these young men on wheels' complained, a letter to *The Times* in 1892:

> To become a cyclist apparently transforms an ordinary young middle-class Englishman into an active member of an unruly mob which it is becoming impossible to ignore. There is no law or order amongst them, and there is no attempt to keep to one side of the road. They come swirling along, sometimes 12 abreast, shouting to each other, or openly making remarks upon any ladies whom they happen to pass.[31]

The ways in which young men used their machines to engage in rowdy and uncontrolled behaviour were far from the only middle-class headache caused by cycling. There was also the question of how normal markers of class distinction could be maintained on a machine which not only had an unwanted tendency to leaving one looking dishevelled, but also took one into new environments where social norms might break down. 'A peculiar trait is shown by a certain class of wheelmen... of fastening themselves on to a passing cyclist, or a party... without the slightest attempt at obtaining permission,' began an article in *Cycling* bemoaning socially unaware cyclists. It went on to state:

> Without wishing to appear in the slightest degree snobbish, there is a clearly defined line of 'class distinction' shown whether mounted upon a cycle or not, and many object to have the company of a stranger a few degrees lower in the social scale thrust upon them.[32]

The challenges that cycling posed to established notions of propriety and respectability were one of its defining features as a pastime. There existed a constant tension between the publications, organisations and individuals who looked to reconcile it with traditional middle-class values, and those who relished the opportunities it created for escaping them. Tellingly, many had a foot in both camps and journeyed across this divide as and when it suited them. The wonderful humour and colour that runs through cycling in this period frequently arises from the conservative social values people sought to apply to the bicycle, and the contrasting free-spirited behaviour that men and women enjoyed when out riding.

And finally, in and among all the debates, passions and controversies generated by cycling, there is also the much more important story of the broader societal changes it helped bring about. Nowhere is this more apparent than in the activities of female cyclists. While tricycling had been established as an acceptably 'feminine' activity, the idea of members of the 'fairer sex' pedalling safeties was initially met with widespread ridicule and criticism. As Lillian Campbell Davidson, a leading female cycling journalist, recalled:

> Cycling women were regarded with a kind of pious horror by society and by the public at large. It was openly said that a woman who mounted a bicycle hopelessly unsexed herself; she was stared at and remarked upon in town. It was supposed that no woman would take so masculine an amusement unless she was fast, unwomanly, and desirous of making herself conspicuous, and accordingly all cycling women had to suffer from the supposition.[33]

Objections to women riding bicycles were centred around how such an activity was seen to conflict with traditional notions of 'femininity'.

For much of the nineteenth century, middle-class attitudes towards women's capabilities and place in the world were strongly tied into discourses of 'separate spheres'. With earning an income presented as a solely male responsibility, women were instead defined by their roles as wives and mothers. Supposedly lacking the drive, energy and intelligence needed to succeed at the workplace, women should instead use their innately gentle and loving natures to create a home in which children could be nurtured and raised, and men refreshed after busy working days. Countless pieces commented on how female cycling conflicted with such understanding, for example:

> Mother's out upon her bike, enjoying of the fun,
> Sister and her beau have gone to take a little run.
> The housemaid and the cook are both a-riding on their wheels;
> And Daddy's in the kitchen a-cooking of the meals.[34]

Of course, understandings of female roles and responsibilities were not absolute, and were subject to continuous debate and negotiation. Certainly by the 1890s notions of women's physical and mental inferiority to men were less extreme than they had been earlier in the century. There was a noticeable increase in the numbers of female clerks, typists and teachers, supported by growing numbers of girls entering secondary education. There was also rising female participation in middle-class sports such as lacrosse, hockey and lawn tennis, as the health benefits which women (and, of course, their potential offspring) derived from moderate forms of exercise became better established.[35]

These developments helped facilitate the huge increase in female cyclists which occurred during the closing years of the nineteenth century. Early dissenting voices were silenced as the numbers of women purchasing and riding machines swelled, with one in three bicycles ordered in Britain in 1895 being for women (this compared to rates of one in fifty just two years previously). The significance of these developments was not lost on contemporaries. There was perhaps no more powerful symbol for modern understandings of 'femininity' than a woman publicly and independently powering a machine which she used to travel where and when she wished. As one writer put it:

> The bicycle is in truth the woman's emancipator. It imparts an open-air freedom and freshness to a life heretofore cribbed, cabined, and confined by convention. The cyclists have collided with the unamiable Mrs Grundy [a voice for conservative opinion in this period] and ridden triumphantly over her prostrate body.[36]

Despite such cheery optimism, it must be noted that at the turn of the twentieth century gender equality was still a distant dream in Britain and indeed all other Western countries. Legal and cultural norms still restricted women's entry into many professions and their access to university-level education, and barred them from the political process. As will be shown in each of the chapters which follow, despite the rapid growth in women pedalling their machines during the 1890s, understanding of appropriately 'womanly' behaviour meant they could not participate in the pastime as fully or freely as their male counterparts.

But at the same time, cycling did undoubtedly play a critical role in challenging older understandings of sexual difference. Away from the home and participating in what had long been a 'masculine' amusement, it was at the forefront of middle-class women's broadening horizons and opportunities. This can be seen in the writings of N.G. Bacon, a leading advocate for women's cycling in this period:

Too long have our lives been narrowed and restricted by the limits of civilisation. The dormant capacities latent in our womanhood have seldom had a chance of being awakened. The cycle allows us, in a perfectly womanly – for I like that word better than 'ladylike', for it betokens something grander, nobler – manner 'to let ourselves go',[37]

The contrast between this and Balfour's earlier assertion are striking. Balfour's was true to his conservative values, celebrating how cycling was now enabling people to find escape and respite from the pressures of the modern world. By contrast, Bacon's looks hopefully forward to the freer and less restrictive society that the bicycle was helping bring into existence. For Balfour, it was the most 'civilising invention in the memory of the present generation', while for Bacon it offered a direct challenge to the 'limits of civilisation'. It is difficult to think of another invention from this or indeed any other period that could so powerfully symbolise two such different and evocative world views.

Of course, it was not just in Britain that the safety bicycle acquired such meanings. In other Western countries it was equally popular and controversial. The focus of the chapters which follow will be on Britain and, most specifically, England. But woven throughout is a flavour of how the British experience compared with that of America, France and other cycling nations from this period, and a sense of how cycling fitted into an increasingly global and interconnected world.

As we shall now see, the history of late nineteenth-century cycling is one which extends far beyond its main protagonist, the bicycle. It offers us a fascinating window into a world both like and unlike our own, in which values and customs that we take for granted today were beginning to take root and alter the face of society. Against the backdrop of these big societal shifts and developments, the pleasures, excitement and humanity of the men and women 'a-wheel' creates a powerful connection to a period which can otherwise feel distant from our own. And as with all good history it allows us to re-evaluate the present, providing a timely reminder of how remarkable and precious an invention the bicycle really is.

1

Fashion, Bikists
and Bloomers

It is difficult to imagine how John Burns, trade unionist and M.P. for Battersea, viewed the events which took place in its well-known park during the summer of 1895. Early on in his time in office, Burns had been instrumental in wresting control of Battersea Park away from its restrictive former owners. Now that it was locally administered, he envisioned it becoming a democratic open space where working-class members of his constituency could enjoy healthy outdoor recreations, safely removed from the temptations of drink or the music hall.

Burns's campaigning was undoubtedly good news for local cyclists. No longer subject to regulation which banned cycling in most of London's other major parks, it was one of the few green, spacious and well-paved parts of the city where they could go and freely pedal their machines. Drawn by such alluring prospects, the numbers of people cycling in Battersea Park swelled during the summer months of 1895.[1]

You would have needed only a quick glance at the men and women riding their bicycles, however, to realise these were not individuals drawn from the social classes Burns had spent his life representing. They not only owned state-of-the-art safeties, but had their machines delivered to them by footmen, who carried them forth from carriages parked around the edge of the park. Quite inadvertently, Burns had played a leading role in the bicycle finding favour with a section of society that had by and large previously ignored it: the aristocracy.

'Batterseason', as it was soon dubbed by *Punch*, saw dukes, duchesses and even royalty (Queen Victoria's youngest daughter Princess Maud made a much remarked upon appearance) circling around the park. Throughout the summer of 1895 crowds were drawn to Battersea to gaze upon 'the cycling exhibition which has been the most curious fashionable rage of the season'.[2] As Jerome K. Jerome, the writer and humourist most famous for his novel *Three Men in a Boat*, recalled in *My Life and Times*:

> In Battersea Park, any morning between eleven and one, all the best blood in England could be seen, solemnly peddling up and down the half-mile drive that runs between the river and the refreshment kiosk... In shady by-paths, elderly countesses, perspiring peers, still at the wobbly stage, battled bravely with the laws of equilibrium... daughters of a hundred Earls might be recognized by the initiated, seated on the gravel, smiling feebly and rubbing their heads.[3]

Jerome was not the only one bemused by the sight of 'perspiring peers' struggling with their machines. 'Riding the bicycle is obviously one of the things which the [upper] classes fail to give their social inferiors "points and a beating",' commented an editorial in the *Cyclists' Touring Club Gazette*. It went on:

> Nearly all the riders sit far too low, carry their hands too high, and are apparently in happy ignorance of what constitutes effective ankle action; indeed, the bulk of the fair sex pedal

with the hollow of the foot, and appear to consider that high French heels were invented to keep the feet from slipping off the pedals altogether… all this savours of 'playing' at cycling.[4]

As these quotations make apparent, the 'upper ten' were much more inclined to treat the bicycle as a toy than as a practical means of loco-motion. After all, a class that could afford the costs of horse-drawn travel had far less need for this new mode of personalised transporta-tion than their 'social inferiors'. What drew them to cycling in Battersea (perhaps excluding the widely attended al fresco 'bicycle breakfasts at eleven') were the newfound opportunities for disporting themselves in front of others. As another journal commented:

It is hardly to be wondered at that one sees so many pretty bicycle costumes in the park and elsewhere. Necessity, which is the mother of invention, has set the wits of all the fash-ionable tailors to work, designing new and *chic* costumes for those wheelwomen who know so well how to dress. Some women go so far as to have specially enamelled bicycles for their different dresses.[5]

Unsurprisingly, most cyclists did not go to the same lengths as their upper-class counterparts when garbing themselves for the pastime. For those focussed on travelling out of cities and into the countryside, 'immense sleeves, pinched waists, painted and powdered faces, high-heeled boots, tremendous hats and a liberal supply of jewellery' were clearly not the order of the day.[6] People for whom the bicycle held a far greater appeal as a practical means of locomotion had little incentive to invest large sums of money on impractical forms of clothing or specially enamelled bicycles.

Nevertheless, even if crowds were not coming out to view them, clothing and appearance were undoubtedly a central preoccupation for all who took up cycling in this period. Cycling was a new pastime in which few had participated previously. For those who now did so, before the question 'Where should I cycle?' came 'What should I wear for my excursions?' As an anonymous writer in the cycling press observed:

> When a person becomes transformed from an ordinary citizen to an enthusiastic cyclist, the question of clothing assumes a different aspect. Ignoring the Scriptural admonition, he begins to grow solicitous as to wherewith he shall be clothed. He recognises, as every sensible cyclist must do, that whatever merits or demerits may appertain to the ordinary civilian dress, its unsuitability for cycling is axiomatically certain.[7]

As we shall see, choosing attire which was both appropriate for riding a bicycle and conformed to contemporary stylistic conventions was no easy matter. The ways in which people sought to navigate these two competing demands not only stimulated a huge consumer market but also generated huge amounts of discussion and controversy. With the aristocracy far from the only group to seize on the bicycle's potential for displaying eye-catching outfits, cycle fashions helped push the pastime even more firmly into the public eye. At the other end of the spectrum, female cyclists who sought more practical forms of cycle-wear provoked debates which extended far beyond cycling into long-standing questions concerning female freedom and emancipation.

As a starting point for these stories, there is no better place to begin than a class of rider that was the first to fully exploit the bicycle's potential for displaying oneself to others: the young clubmen who banded together on their ordinaries.

4 Print of aristocratic riders in Hyde Park (1896). As other London parks removed restrictions previously placed on cyclists, upper-class riders began to venture away from Battersea.

5 Members of a high-wheel club displaying their machines and uniforms during a run (date unspecified).

'What ho, she bumps!'

Writing in the *Manchester Guardian* in 1893, long-time cycling journalist R.J. Mecredy reflected on an issue that had dogged cycling since its early years as a pastime. Few, he presumed, would disagree with his assertion that:

> Cycling is the ideal recreation for the business or professional man. It completely occupies his mind for the time being. The change of scene, the incidents of the road, the attention necessary to bestow on one's machine – all unite to attract his attention, and he will have hardly have travelled a mile before the tired, numb feeling in the head will have given place to one of exhilaration.[8]

Given these apparent benefits, his article 'Cycling for Brain-Workers' asked why cycling had not then achieved universal popularity among this section of society. The answers, he stated,

> ...are prejudice and snobbishness. Staid business and professional men cannot reconcile themselves to the fancied indignity of straddling a two-wheel affair like the bicycle, and they cannot recognise that it is both healthy and pleasant... Snobbishness is even a worse enemy to the spread of cycling. It is the recreation of the masses, and the bicycle is the poor man's horse. Hence the classes look upon it with disdain.[9]

Mecredy's sentiments were echoed the following year by *Cycling*: 'it is the "horsey" instinct we take it that has been the largest factor in practically barring the use of the cycle in the upper ranks of society'.[10]

It would not be until 1895 and the Battersea Park phenomenon that the aristocracy would finally take to cycling in large numbers, and the bicycle lose its reputation as 'the poor man's horse'.

In the years leading up to 1895, cycling was an activity which carried an uncertain set of social connotations. Those lucky enough to possess the money and free time to cycle were undoubtedly well-off when compared with the majority who lived and worked in Britain. Most cyclists were relatively prosperous 'brain workers' such as clerks, accountants and tradesmen, a step or two further down the social ladder than the 'staid business and professional men' described by Mecredy. Because of the lack of patronage bestowed on the bicycle by the upper ranks of society, however, it could still be presented as a 'low, vulgar pastime… not fit for gentlefolk'.[11]

The ambiguous class associations attached to the bicycle in its early years do much to explain the great lengths to which ordinary riders went to affirm their social status. A significant proportion of high-wheeled cyclists were members of clubs, attracted not only by the protection and sociability they offered, but also the 'social standing in the bicycling world' which came with membership.[12] Clubs modelled themselves as select organisations, keen to distinguish themselves from 'rank and file' ordinary riders, otherwise known as the 'great unattached'. Writing in the 1890s, long-time clubman Charles Sisley recalled:

> To be a member of the 'Great Unattached' was considered almost a disgrace, leading others to assume that the rider was not of sufficient respectability to become a club member. The club uniform was as it were a masonic sign, which acted as an introduction on the road. A rider attired in one of the well-known uniforms… was considered a desirable acquaintance when met on a touring exhibition… Many a score of pleasant

evenings have I spent in the company of perfect strangers…
whose sole introduction was a recognised club uniform.[13]

Club uniforms thus acted as distinctive status symbols for their owners.[14]
The cost of these outfits, typically around thirty shillings, meant they
were a clear representation of their wearer's relative wealth and prosperity. Their unique design also ensured that members of clubs were
sharply differentiated from other ordinary riders. Uniforms managed
to fill what was a highly challenging brief – making the young men
atop high-wheeled machines even more noticeable figures.

Flamboyant cavalry style outfits were the norm for most clubs,
consisting of a double-breasted, close-fitting patrol jacket with a
straight waistcoat, breeches or knickerbockers tucked into stockings,
and a pillbox hat or cadet cap.[15] From this typical starting point they
experimented with different colours and military-inspired trimmings
to create a costume exclusive to their institution. One can well imagine
that the Preston Bicycle Club, in their dashing ruby red cord breeches,
brown jackets, stockings and jockey caps, would have made a striking
sight as they pedalled out into the Lancashire countryside.[16] At the
beginning of his official history of the Anfield Bicycle Club, Frank
Marriott recollected how during cycling's early years he had seen

> … riding in the principal street of the city, a gentleman wear-
> ing crimson plush breeches and a coat of bright blue, braided
> with yellow, with a good many brass buttons. He was *not* an
> Anfielder. We wore black, black from head to foot; there was a
> little 'hussar' braiding on the jacket, and the braid was black.
> We resisted the temptation to a cap-badge of skull and cross-
> bones in white; but we had stripes of royal blue on the black
> scarf… Our dress was decent, sober and correct.[17]

It is not surprising that clubmen took a smug satisfaction in their uniforms, used as a means of distinguishing themselves from the 'great unattached'. As we shall see, this was far from the only military-inspired aspect of early club life. Weekend runs were usually undertaken in a highly disciplined manner, as members travelled in orderly ranks behind a recognised 'Captain'. The importance clubs attached to conveying a 'respectable' image of themselves can be seen in a piece in the Polytechnic Cycling Club's magazine *Home Tidings*, which exhorted members to

> … make the Polytechnic C.C. notorious for their big musters, smart appearance and everything that pertains to the reputation of the Club… the uniform (grey jacket and breeches, with straw hat, or black cricket cap) is one of the neatest in London.[18]

Appearing 'decent, sober and correct' was, however, far from the sole motivation for members investing so much time and effort in choosing attire for their weekend excursions. A key reason why they spent a considerable portion of their income on bicycles and uniforms was the subsequent opportunities for showing themselves off during Saturday runs. After long weeks spent working under the instructions of superiors, high-wheel riding was quite unique in allowing young men, in their own minds at least, to take on exotic, devil-may-care personas. This is evident in Flora Thompson's *Lark Rise to Candleford*, where she recalled:

> Members of the earliest cycling clubs had a great sense of their own importance, and dressed up to their part in a uniform composed of a tight navy knicker-bocker suit with

red or yellow braided coat and a small navy pill-box cap embroidered with their club badge. The leader carried a bugle suspended on a coloured cord from his shoulder.[19]

Thompson described how, after arriving at Candleford, clubmen would send 'facetious telegraphs' home describing supposed accidents and difficulties they had experienced over the course of a ride, each with the aim of proving himself to be 'a regular devil of a fellow'. She remembered them as 'townsmen out for a lark':

> After partaking of refreshment at the hotel they would play leap-frog or kick an old tin on the village green. They had a lingo of their own. Quite common things, according to them, were 'scrumptious', or 'awfully good', or 'awfully rotten', or just 'bally awful'. Cigarettes they called 'fags'; their bicycles their 'mounts', or 'my machine' or 'my trusty steed'; the Candleford Green people they alluded to as 'the natives'. Laura was addressed by them as 'fair damsel', and their favourite ejaculation was 'What ho!' or 'What ho, she bumps!'[20]

Thompson was far from the only person to view high-wheeled clubmen as peculiar and amusing rather than dashing and debonair. Looking back on this period at the turn of the twentieth century, the *Daily Graphic* recalled that the early cyclist was

> ... popularly regarded as an eccentric being, attired, as a rule, in a braided jacket and jockey cap, which gave him the appearance of a circus rider, and probably did much to retard the popularity which was bound to come later on.[21]

This was undoubtedly an overstatement. When looking at what put people off cycling in its primary years, the writer would have been better off focussing on 'headers' and the inherently dangerous nature of high-wheeled riding, rather than the costumes worn by cyclists. Nevertheless, it is telling that as ordinaries were abandoned in favour of safeties, the clothing associated with early cycling also began to disappear. Although (or less generously, because) cyclists remained highly image conscious, as early as the middle of the 1890s tight knee-length breeches and braided jackets were being viewed as outdated relics from the past. Such outfits were incompatible with a pastime which now shed its cultish reputation and snowballed into 'the craze of the hour'.

The latest fashion

During the closing years of the nineteenth century, cycling was a social phenomenon. It was sung about in music halls, subject to extensive newspaper coverage and debated in the House of Commons. Cab drivers, stable keepers and others who worked with horses and carriages complained that their customer base was disappearing, while in the fashion industry there were real worries about business declining calamitously. One journal complained:

> Women nowadays have not the time as hitherto to devote to the daily etceteras of their toilet, and spend their dress allowance on their cycling rig-out instead, and any extra pocket-money they possess goes in purchasing new machines.[22]

There were some grounds for such grumblings. From the mid-1890s the numbers investing their time and money in the new pastime increased dramatically. Between 1895 and 1896 the number of new members joining the Cyclists' Touring Club nearly quadrupled to 21,000, and the organisation continued to expand rapidly until the end of the decade.[23] This was reflected in the rapid expansion of the British cycling industry. In London alone, the number of bicycle manufacturers was calculated to have increased from 152 in 1894 to 390 in 1897, while nationally the amount of money invested in large joint stock companies involved in the cycle trade increased from £1 million to close to £20 million in the same period.[24]

When investigating what caused such rapid developments it is easy to focus on the invention of the safety bicycle and pneumatic tyres. These advances were unquestionably crucial in cycling becoming a far more enjoyable and attractive pastime than it had been previously. The bicycle's newfound practicality was, however, far from the only factor driving its widespread popularity. Quite suddenly, it also acquired a reputation as highly fashionable. As a leading society paper put it:

> The cycle is not to be confused with the 'bike'. There is a distinction, if not a difference, between them, and the distinction belongs exclusively to the former. The cycle was a thing of utility, the sight of it at one's door carried immediate suggestions that the man about the taxes, or the water-pipes, or the gas had come; the sight of the 'bike' in the same position thrills one with the thought that one's most fashionable acquaintance is paying a possibly long-deferred call.[25]

We have already touched upon a key driving factor in these changes: the sudden uptake of cycling among members of the aristocracy in the spring of 1895. As newspapers photographed and recorded the various society figures taking to their cycles, it was soon established as a chic and highly desirable new pastime, giving 'confidence to the timid that they, too, might go pottering in the parks without losing caste'.[26] The patronage cycling enjoyed among distinguished female figures also settled the question of 'whether ladies can with propriety ride the bicycle'. Constance Everett-Green, a prominent female cycle journalist, later recalled:

> It would hardly be too much to say that in April of 1895 one was considered eccentric for riding a bicycle, whilst by the end of June eccentricity rested with those who did not ride.[27]

As the cycle market boomed during the middle years of the decade, however, other forces helped entrench cycling's modish new reputation. Most notable of these was something which few of us might associate with this period in history: sophisticated forms of advertising. In a hugely lucrative market, firms were prepared to invest significant sums in promotion and marketing. The international nature of cycling – which was undergoing similar surges in popularity in America, France and other Western countries – meant British firms not only had to compete with each other, but also with foreign manufacturers looking to sell their machines overseas. After American companies began to establish themselves in Britain in 1896, a correspondent wrote to *Cycling*:

> The American manufacturers have deserved success by reason of their splendid enterprise and judicious advertising… British manufacturers must not let the Americans eclipse them in enterprise or advertising schemes.[28]

The 'splendid enterprise and judicious advertising' employed by American manufacturers was nowhere more apparent than in the figure of Albert Augustus Pope. One of the first American businessmen to import British bicycles in 1878, Pope then used these as inspiration for his own brand of 'Columbia' high-wheelers. By the 1890s Columbia safeties were being sold all over the world, with Pope the head of a highly successful international company. At the forefront of bicycle sales and marketing, he utilised all manner of methods for selling his machines, from sponsoring leading riders to developing a national network of sales agents. During an interview in 1895 when Pope was asked about his best employee, he commented:

> He is the most faithful fellow in the world. He has been in my employ for seventeen years, yet he has never even asked for a holiday. He works both day and night, is never asleep or intoxicated, and though I pay him more than $250,000 a year, I consider that he costs me nothing. His name is Advertisement.[29]

For Pope and other manufacturers, a significant portion of their advertising budgets went towards publicity posters, often produced by recognised artists such as Toulouse-Lautrec and Maxfield Parrish. Thus one of the most noticeable legacies of the 'craze' are the eye-catching, colourful and visually striking posters which were commissioned on a global scale. As well as taking inspiration from the newly established Art Nouveau movement, posters frequently used female cyclists as 'the goddesses of this revolutionary conquest of liberty'.[30] The use of an innovative and original art-form to visualise the most current of social spectacles – a woman riding a bicycle – asserted cycling's place at the forefront of modern life.

Posters also encouraged a view of cycle riding as a comfortable and enjoyable pastime, a source of individual freedom and emancipation. Cycle companies are recognised to have been some of the first businesses to move away from using large amounts of text to describe their product's characteristics, focussing instead on the enjoyment and gratification it would bring to its owner.[31] Their marketing efforts offered forms of wish fulfilment we associate with modern advertising: purchase our brand of bicycle and be as happy, sophisticated and graceful as the person shown in the poster.

For those of us familiar with such selling techniques, it comes as little surprise that such a vision did not always correspond to reality. More experienced riders, familiar with dodgy roads, punctures and the British climate, were certainly not wholly convinced by these new forms of publicity. Jerome K. Jerome could have been commenting on any number of adverts when he stated:

> In ordinary cases the object of the artist is to convince the hesitating neophyte that the sport of bicycling consists in sitting on a luxurious saddle, and being moved rapidly in the direction you wish to go by unseen heavenly powers.
>
> Generally speaking, the rider is a lady, and then one feels that, for perfect bodily rest combined with entire freedom from mental anxiety, slumber upon a water-bed cannot compare with bicycle-riding upon a hilly road. No fairy travelling on a summer cloud could take things more easily than does the bicycle girl, according to the poster.[32]

As those who took up the pastime quickly discovered, there could be a considerable gap between your experience of cycling and the one you had been sold by the manufacturers. A new female rider bemoaned the fact that when cycling:

> One perspires so horribly, and after half an hour's ride one gets into a dreadful state. My cousin Anna simply melts when she is riding. I always take a little powder-box and a pair of tongs and a spirit lamp to curl my fringe, but it is very difficult to use them when gentlemen are present, for that makes such a fuss, and they might laugh at one. I am always getting bruises, too, and hurting myself. I hope the fashion will soon die out.[33]

By no means enamoured with the new pastime, this lady cyclist was a prime example of what those in the cycling press widely termed a 'bikist'. This was a label used to describe those individuals who had not previously cycled but had been drawn in following its newfound fashionable reputation. The main concern of 'bikists' was not the new places they could explore and travel to on their wheels, but rather how they could most appropriately partake in the new fashion.

As evidenced by the 'dreadful state' the spirit-lamp-wielding cyclist got into, this was not always something they managed to achieve. It is easy to forget that those who started cycling during the 1890s had to achieve in adulthood what we all learnt as children: how to ride a bicycle. Staying abreast with the latest 'craze' therefore required you to follow an often painful and bruising learning curve as you attempted to gain mastery over your machine.

In the absence of supportive parents, there were many individuals who (for a decent fee) could see you through this difficult process. The growth in those looking to take up cycling saw a corresponding increase in 'cycle instructors' who operated in parks or large indoor spaces, such as halls or gymnasiums, which were adapted into special-ist 'cycling schools'. Prices ranged from about two shillings for a single lesson to ten shillings (about £35 in today's money) for 'full tuition', with additional fees if you needed to hire a machine.

In their pupils' early stages of learning to remain upright, the main purpose of instructors was to offer moral and physical support. The most common teaching mechanism was relatively simple, consisting of putting an arm around the side of the cyclist they were mentoring, and guiding them forward until they felt confident to pedal by themselves. After passing through this stage, they then taught their students the valuable skill of how to 'correctly' disport themselves when riding out in public. As a leading female fashion journal commented:

> The art of managing a bicycle is becoming, if it has not already become, a necessary part of a young lady's educa-tion… To-day, we find throughout the length and breadth of the country, cycling schools, which teach ladies not only to ride… but to ride correctly and well – that is to say with good seat and perfect command of the machine, with a carriage that is neither bold nor timid, neither formal nor rigid, nor yet careless. To teach a lady to comport herself on a bicycle as she has been taught to comport herself on foot, the cycling word is indebted to these schools.[34]

Once you had learnt to ride your machine 'correctly and well', there was then the question of where best to participate in the new fashion. Following the example of the aristocracy, most bikists took themselves to large inner-city parks. These were not only conspicuous spaces where they could display themselves to considerable numbers of people, but their flat, well-paved roads also lent themselves to gentle forms of exertion. This allowed cycling school graduates to show off their 'good seat and perfect command of the machine', while also exhibiting outfits which would not survive longer journeys into the countryside.

The entry of this new class of riders into the ranks of cyclists was good news for some more experienced riders. For a few years, young men who were experienced cyclists found gainful employment as cycle instructors teaching trainee bikists. Jerome K. Jerome, evidently finding much inspiration from all things cycle-related, commented on this aspect of the craze:

> I have serious thoughts of abandoning literature and journalism, and becoming a bicycle instructor... I am told that good-looking young men, possessed of agreeable manner and a knowledge of flirtation, receive a good deal in the way of tips... Promenading a park, with my arm round a pretty girl's waist, is, I feel, the vocation for which nature intended me.[35]

Most seasoned cyclists, however, extended a far cooler welcome towards the fashionable riders who became increasingly prominent during the 1890s. In his cycling novel, *The Potterers Club*, George Lacy Hillier mocked

> ... the bikist just emerging from the riding school and park-promenade, clad in daintily coloured flannels; trousers, de rigueur in this cult, clipped round his ankles, brown boots tightly laced, soft felt hat, and – hall mark of the true bikist – high linen collar with neat tie well pulled up.[36]

Members of the cycling press also directed considerable scorn at the female members of this new class of cyclist. A disapproving article in *Cycling* described the 'giddy girl bikist' as one who

> ... rides a wheel, but really you can't call her a lady cyclist. Does she ever go for a run in the country? Does she ever use her wheel save Saturdays, and perhaps Sundays, and other evenings after 'the dear boys' are home from business? Not she; as she herself will tell you, she is 'not such a silly'. In the parks and along those remarkable strips of suburban roadway – known further East as 'Monkey Parades'... the 'giddy girl bikist' flits here and there dispensing bows, smiles and half glances as she goes.[37]

It is unquestionable that for many living through the 1890s the bicycle represented less a liberating new form of personalised transportation, and more a stylish and novel plaything. This is not to say they did not find cycling an enjoyable or exciting new experience, but rather that their enthusiasm came from participating in the latest craze and the various opportunities cycling created for showing off to others. The writer of the piece above, titled 'Cycling or Posing. Which?', went on to describe the activities of 'immaculately attired cherubs' who

> ... mount their cycles, and are thus able to exhibit themselves to more admirers than would be possible if they walked; they, as it were, get upon their triumphal cars and are carried round in state. Riding with high handles and low gears, cycling disarranges their elaborate 'get-up' even less than does walking.[38]

Of course 'bikists' were far from the only class of cyclist to emerge in great numbers during the 1890s. For every one of them there were countless others who, after long working weeks in the hearts of cities, could not wait to mount their 'glittering wheels' and find release and escape in journeys out into the countryside. For these people there existed a desire for something which was of far less consequence to fashion-conscious park cyclists, or indeed the young men who had looked to impress 'fair damsels' atop their ordinaries – and that was practicality.

'On space-devouring thoughts intent'

Cycling's transformation into a comfortable and everyday activity following the invention of the safety bicycle and pneumatic tyre is clearly evidenced in the clothing of male cyclists from this period. Outfits for a pastime that now appealed to men of a wide range of ages and athletic abilities kept pace with demands for freer and more functional apparel. Breeches that hugged the top of the thigh were done away with in favour of baggier knickerbocker trousers, which would be tucked into hoses or stockings to prevent them catching on the chain. Restrictive, military-style outer garments were also replaced with looser-fitting Norfolk jackets, which towards the end of the nineteenth century were becoming increasingly used in other outdoor sports such as golf and shooting.

Compared to their predecessors, 'cycling suits' also made far greater use of woollen materials. The breathability, warmth and water-resistance of wool meant it was commonly viewed as the most appropriate material for long-distance cycle rides. F.T. Bidlake, a famous racing cyclist from this period, commented that for the average rider

… it is not enough that his shirt be woollen, every portion of his clothing should all be of the same material. Stiffenings, linings, pockets all pure wool. It may sound faddy, but it is the secret of comfort. Woollen garments are the coolest in the heat, and the warmest in the cold, a wool-clad cyclist can stand a thorough soaking with impunity, and his clothes will not cling to him with that chilling and deadly embrace of cotton or linen.[39]

For Bidlake and other more hardened cyclists, woollen suits were the gold standard when it came to practical forms of cycle wear. After scoffing at so-called 'bikists' and 'butterfly riders', George Lacy Hillier was much more complimentary about the 'steady tourist'

… garbed in sound and workmanlike fashion from head to foot: a serviceable suit, well made and well fitted, a buttoned flap over every pocket to retain the content and exclude the wet; an easy Norfolk jacket… stout stockings, light spats and business-like shoes… Here is the hard road rider, rather dusty, a trifle unkempt, clad in loosely-fitting garments.[40]

It would be a mistake, however, to view cycle suits as being simply practical outfits whose appeal rested solely with 'hard road riders'. Admittedly, they were less ostentatious than the clothes worn by members of early cycle clubs or those who promenaded in parks. But at the same time, those wearing this new mode of apparel sought to project an image of themselves which did far more than avoid Lacy Hillier's warning that a cyclist clad in 'inappropriate costume… proclaims himself as either an ignoramus or a novice'.[41]

Continuing older traditions, wearing a suit outwardly affirmed your class status to those you encountered during a ride. The cost of a full-price outfit, consisting of shoes, knickerbockers, stockings, shirt, jacket and cap, could equal the price of a new machine, meaning it acted as a clear marker of its owner's wealth and social standing. Furthermore, modelling this attire during weekend excursions showcased one's conformity to the middle-class belief that participating in leisure activities was no excuse for abandoning smart and well turned out clothing. W. MacQueen Pope, recalling his middle-class London childhood at the turn of the twentieth century, reminisced:

> One dressed for a holiday then, one did not undress. One kept up one's station and showed one's position in life as much by the Sea as in Queens Road, Finsbury Park... Your attire could be less formal, slightly more colourful, a little easier... A man could wear flannel shirts and a blazer... his straw boater, but he did that in town; he might, if middle aged, wear a Panama, or sometimes he wore a cricket cap perched rakishly on the back of his head.[42]

Advertisements for cycle suits also make it clear that they were designed not only as 'respectable' attire but as stylish and sophisticated pieces of clothing. The popularity of cycling during the 1890s, as well as of other sports such as rowing and golfing, meant that looser-fitting leisurewear was marketed not only as being suitable for outdoor recreations, but as a fashion in itself. As such, unlike earlier club uniforms, suits were very much in line with contemporary style trends.[43]

This can be seen in a catalogue of cycling garments produced by London-based Holding and Son Tailors. At a time when those who belonged to the Cyclists' Touring Club could distinguish themselves as C.T.C. members by purchasing an official uniform, Holding and Sons

Copyright

THE "TAILOR & CUTTER" FASHIONS, 1902

PUBLISHED BY THE JOHN WILLIAMSON Cº LIMITED · LONDON. W. C.

6 Men's cycling suit (and dapper cyclist) included in the 'Tailor and Cutter' fashions for 1902.

acted as the club's 'official outfitters'. Unlike the clothes worn by early institutions, the C.T.C.'s uniform for 1897 consisted of:

> Dark green Devonshire serge Norfolk jacket or tunic, knickerbockers or breeches... helmet of the same material, green stockings and white gloves.[44]

After promising prospective buyers that 'we turn out best work only in the latest styles' and selling themselves as giving 'C.T.C. members who live away from fashionable centres an opportunity of getting high-class clothes made in the best West-end way', the catalogue stated:

> Though engaged in the best class of West-end Tailoring, employing skilled and highly-paid labour, we are still able, as official outfitters to the C.T.C., to execute, with superior excellence as to work, styles and fits, orders in the genuine Club materials at list prices.
>
> Twenty-two years riding has shown us the best principles on which to cut, make, and fit cycling garments; whilst many years' experience as head cutters in firms of the highest repute, is a guarantee of satisfaction being given.
>
> Annexed measuring form filled in and returned with One Guinea, will ensure a pair of stylish trousers, in Club or other material, being sent to any address in three days, or less if desired. Patterns may be selected from our stock of 400 samples, and sent by return of post.[45]

Although not all who joined the C.T.C. undertook long-distance tours and rides, it is striking that the club's official tailor gave more attention to how their outfits were informed by 'the latest styles' than their actual suitability for cycling. Even for more seasoned riders, it was felt that they would be just as interested in the 400 patterns that could be used on 'a pair of stylish trousers' as in the strength, lightness and comfort of the garments they purchased. As 'sound and workmanlike' as cycling suits may have appeared to some, their purpose quite clearly extended beyond aiding their owner's weekend excursions to the country.

This is further evidenced by the fact that when it came to taking exercise on a bicycle, these were outfits which came with their fair share of limitations. As breathable as wool might be, there are good reasons why we no longer cycle in trousers, shirts and jackets on hot summer days. While some riders made use of 'cellular' shirts and caps, which were made with a breathable, open form of cotton, it remained the case that 'excessive sweating is the prevailing bane of most cyclists'.[46]

Faced with this situation, the seemingly obvious thing to do was to dispense with heavy garments in favour of clothes which let your body breathe freely. Some long-distance riders did just this, and dressed themselves in much lighter shorts, tights, shirts and vests, to some extent resembling the Lycra outfits worn by racing cyclists today. For men to display bare parts of their anatomy on roads and other public spaces, however, was in direct conflict with middle-class notions of propriety and decency. Writing in 1897, one commentator warned that the young man who took up cycling might

> ... develop into that terrible creature 'the scorcher', who, absolutely unmindful of the beauties of the scenery through which he passes, neglects, in his one idea of pace, the clothes and customs of a respectable individual.

The author of this piece, Gilbert Floyd, went on:

> He will affect strange garments, sad as to colour, and roomy in cut, whilst his shoes must be stamped from one piece of leather to meet his depraved taste for simplicity and lightness. His only adornments will be his club badge and the silk handkerchief, in which are embodied his club colours, which he wraps loosely round his neck, leaving the ends to flutter pennon-like in the breeze, as he urges his wild career, on space-devouring thoughts intent.[47]

Floyd was not alone in criticising the clothes that 'scorchers' (whom we shall encounter again in the next chapter) chose to wear when sprinting along main roads. An article in *Cycling*, titled 'Racing Costume on the Road', commented:

> When we see, as we too often do, whole batches of men coming out on the highways in the most gossamer-like tights, or even ordinary path-racing drawers, supplemented by their cutaway jerseys or vests, one questions the desirability of the proceedings; and when one catches sight of flying road riders, clothed above the waist-band with an unbuttoned jacket *only*... the unprejudiced observer is forced to admit that the limit line between appropriate clothing for the occasion, and a regard for proper decency, has been passed.[48]

Faced with such criticisms, it is not surprising that most serious male cyclists stuck to their knickerbocker trousers and smart Norfolk jackets. Although far from ideal when riding in the heat, these were outfits which managed the difficult task of meeting the competing demands of fashion, respectability and practicality. The appeal they enjoyed among male cyclists preoccupied with their appearance awheel can be seen in H.G. Wells's Mr Kipps who, after learning he is to inherit a fortune, is not just thrilled by the idea of purchasing a new machine but that 'he might buy a bicycle and a cyclist suit'.[49]

Moreover, before feeling too much sympathy for the male cyclists who sweated at the hands of middle-class notions of 'respectability', we should spare a thought for their female companions who, during this period, had to contend with what were far more restrictive understandings of what constituted 'indecent' cycle clothing.

'One of these here New Women'

We're all doing it now. It's glorious; the nearest approach to wings permitted to men and women here below. Intoxicating!... And it's transforming clothes. Short jackets and cloth caps are coming in. Bustles are no more. And, my dear – *bloomers* are seen in the land!... We're all getting most thrillingly *fin-de-siècle*![50]

Rose Macaulay's *Told by an Idiot* reminds us that those women who took to cycling in the 1890s did not only find the physical experience of riding a bicycle exhilarating. Just as thrilling could be the impact it was having on female clothing. After society cyclists resolved the issue of whether it was 'becoming' for women to ride bicycles, attention swiftly turned to the dress they wore when out riding. And among the many debates which raged around, female cycle clothing was one almost impossibly controversial issue: what exactly should they wear on their legs?

The standard response to this was simple: skirts. Throughout the nineteenth century, ideas of 'respectability' had demanded that women appearing in public should keep their legs safely hidden beneath layers of clothing. The drop-framed bicycle, now widely used for city cycling by both sexes, was initially invented in the late 1880s for this very reason. The absence of a cross-bar across the top of the machine allowed women to cycle in skirts, meaning they quickly became marketed as ladies' machines, with diamond-frame models advertised exclusively for men.[51]

But although drop-framed bicycles accommodated skirted riders, they could do little to prevent the risks that accompanied cycling in this mode of attire. While machines might be fitted with chain guards, flowing garments which extended close to the ground had an unwelcome but not unsurprising tendency to get caught up in the chain and pedals. As one female cyclist, Helena Swanwick, succinctly put it:

> It is an unpleasant experience to be hurled on to stone setts and find that one's skirt has been so tightly wound round the pedal that one cannot get up enough to unwind it.[52]

Though shorter outfits helped counter this problem, these were prone to cause other 'unpleasant experiences'. Skirts cut around the ankle had an embarrassing inclination to fly upwards when travelling down hills or in windy conditions, providing a very public display of their owner's lower limbs. One concerned clergyman wrote in to the *Cyclists' Touring Club Gazette* arguing against women cycling in skirts, as this had resulted in

> ... myself and the group of idle lads at the street corner seeing what none of the male sex ought to be allowed to see,

and what every woman with any pretention to modesty takes great care to avoid being seen.[53]

Considering these difficulties, it was perfectly understandable that some women sought more practical forms of clothing, which soon became widely known as 'rational dress'. In its most commented upon form, this consisted of a smart shirt, blouse or jacket, combined with knickerbocker trousers and stockings. Such an outfit not only combatted the dangers which came with skirts, it also offered less resistance to headwinds and gave freer movement to the legs when turning the pedals. Furthermore, knickerbockers allowed women to ride cross-framed machines which were structurally stronger and about ten pounds lighter than drop-framed models.[54]

Resembling the cycling suits worn by men, rational cycling dress created a new chapter in a long-standing movement for more practical forms of women's clothing. The American Amelia Bloomer had achieved international prominence in the mid-nineteenth century by promoting baggy trousers like those worn by Turkish women (soon widely known as 'bloomers'), which she combined with short knee-length skirts. Bloomer edited the first newspaper written exclusively for women, *The Lily*, and used this platform to argue:

> The costume of women should be suited to her wants and necessities. It should conduce at once to her health, comfort and usefulness; and, while it should not fail also to conduce to her personal adornment, it should make that end of secondary importance.[55]

Building on Bloomer's cause, the Rational Dress Society was founded in Britain in 1881 under the Presidency of Viscountess Florence

Harberton. Highly critical of the health risks and physical constraints imposed upon women by tight corsets and long dragging skirts, it endorsed less restrictive garments such as the divided skirt, similar in design to modern culottes. Its gazette, briefly published in the late 1880s, furthered Bloomer's arguments:

> What can be the true state of intelligence of a creature which deliberately loads itself with quantities of useless material around its legs, in spite of discomfort and danger, without any object in view, beyond the abject copying of one another?[56]

Despite such protestations, however, the lack of women willing to abandon skirts for lighter forms of clothing meant that by the mid-1890s the movement towards rational dress had lost significant momentum. Within this context, growing numbers of female cyclists not only helped revive it, they also helped propel it to levels of publicity and popular attention that far surpassed anything it had achieved previously.

This was partly a consequence of the fact that, compared to bloomers or the divided skirts promoted by Harberton, rational cycle dress possessed a much broader appeal. It was designed to enhance a highly popular activity, and commentators were quick to note the uptake it enjoyed among female cyclists. One correspondent to the C.T.C.'s monthly gazette in January 1896 noted that 'in my excursions to the outskirts of London I have latterly noted with satisfaction an increasing number of Rational dresses', an observation which was echoed by others during the early years of the 'craze'.[57]

But while more women may now have been wearing rational clothing, it must be noted that only a very small percentage of female cyclists were prepared to expose knickerbocker trousers to the public. Their place in the popular imagination was driven less by first-person sightings, more by their powerful symbolism. A lady cyclist, independently enjoying the freedoms and possibilities created by cycling, was a well-established image for female emancipation. Rational dress had for decades previously challenged social norms which prevented women from enjoying the same opportunities and freedoms as men. Bring the two together and you could not have asked for a more compelling representation of what contemporaries widely dubbed the 'New Woman'.

First coined in 1894, this term emerged at a time when middle-class women in Europe and the United States were experiencing greater physical independence, legal rights, and educational and employment opportunities than their predecessors. It was used to describe those who were most actively seizing these newfound opportunities to lead lives which challenged older Victorian ideals of 'womanhood'. In the words of Claire Simpson:

> Although there were numerous interpretations about what constituted the New Woman, the distinguishing characteristics were her independent spirit and her athletic zeal. The New Woman played sport, wore her skirts above her ankles, loosened her corsets, wanted a good education, expected to marry and have children, but also wanted a life beyond her home, maybe even a career... [she] stood for political, social and economic equality.[58]

For some more 'modern' voices from this period, this was something to be celebrated. In *The Wheels of Chance*, after an encounter with a female cyclist in rationals during his cycle tour (a meeting originally caused by him crashing his bicycle at the sight of her), H.G. Wells's Mr Hoopdriver reflects:

How fine she had looked, flushed with the exertion of riding, breathing a little fast, but elastic and alive! Talk about your ladylike, homekeeping girls with complexions like cold veal!... He speculated what manner of girl she might be. Probably she was one of these here New Women. He had a persuasion the cult had been maligned... His mind came round and dwelt some time on her physical self. Rational dress didn't look a bit unwomanly.[59]

There were few others, however, who came to Hoopdriver's conclusion. By directly challenging traditional understandings of 'femininity', women cycling in knickerbocker trousers received considerable amounts of ridicule from both the press and the wider public. One newspaper reported a 'lady bicyclist' describing how, when travelling through a large town at lunchtime,

> ... thirteen persons saluted me with the polite command to 'Git yer 'air cut!'... A ragged urchin ran alongside for some distance, and asked: 'Could yer oblige us with a match, guv'nor?' A barber farther down the road went one better by standing on his step and enquiring: 'Shave, sir?'... while an elderly lady imparted the information that I was 'a forward young minx!' One man – how I thanked that man – doffed his oily cap and exclaimed: 'Bravo! I likes yer pluck!' In spite of the attention my appearance excited, several acquaintances passed me without notice.[60]

Quite often, mockery extended into outright aggression, and in some rare cases this took the form of physical violence. One rationally clad cyclist, for instance, was hit by a meat hook thrown at her when riding through Kilburn in London. More common was hostile verbal abuse, far less good natured than the comments described above. Kitty Jane Buckham summarised these types of intimidation when she wrote to her brother:

> It certainly can't be worse to ride in Oxford than in London, especially London suburbs. It's awful – one wants nerves of iron. I don't wonder now in the least so many women having given up the R.D. [Rational Dress] Costume and returned to skirts. The shouts and yells of the children deafen one, the women shriek with laughter or groan and hiss and all sorts of remarks are shouted at one, occasionally some not fit for publication. One needs to be very brave to stand all that.[61]

Fear of being on the receiving end of such antagonism was undoubtedly a key factor in many women choosing to steer clear of rational dress. It was with the likes of Buckham in mind that a female reporter for *Jackson's Oxford Journal* commented:

> I envy the women who possess that sort of courage, but I could not copy them to save my life, I verily believe, unless it became general to ride in knickerbockers... A skirt is intrinsically feminine, though it is idiotically irrational for cycling.[62]

Faced with such hostility, it was natural for women who cycled in rational costume to band together. The 1890s saw the formation of several influential organisations, such as the Lady Cyclists' Association and the Rational Dress League. These were institutions which not only provided mutual support and assistance to members when out riding together. They also aimed to fulfil a much more ambitious brief by carrying on the work of Bloomer and Harberton to promote reforms in the dress worn by women.

'The battle of the bloomers'

The Lady Cyclists' Association and Rational Dress League were certainly well placed to build on the earlier work of Viscountess Harberton: she was a leading member of both. Now in her fifties, she remained committed to modernising women's clothing so it was less restricting, more hygienic and better suited to everyday activities. To this end she played a prominent role, particularly for the Rational Dress League, in supporting their efforts to gain public acceptance for 'the wearing by women of some form of bifurcated garment, especially for such active recreations as cycling, tennis, golf, and other athletic exercises, walking tours, housework and business purposes'.[63]

The most notable of Harberton's endeavours took place in 1899 when she brought a widely publicised court case for refused service against the landlady of a Surrey hotel. The landlady in question had denied her entry to the hotel's coffee room and instead asked her to take luncheon in the bar parlour, for fear of the scandal that her clothing might create. Harberton, supported by the Cyclists' Touring Club, was the central figure in a lawsuit that became known as 'the battle of the bloomers' and received extensive media coverage.[64] While the jury found in favour of the landlady, Harberton succeeded in what was the main motivation behind the suit – raising awareness of the hostility faced by women who cycled in rational clothing.

To publicly shine a light on this animosity, however, was one thing. The real challenge facing Harberton and her supporters was how they

7 Viscountess Harberton (on right) photographed in rational dress costume alongside Frances Heron-Maxwell, secretary for the Rational Dress League (date unspecified).

could best tackle and reduce it. When confronted with such widespread scorn and aggression, how could you reform public opinion to accept your right to cycle in more practical clothing?

Luckily, answers were at hand. After all, only a few years previously female cycling itself had been viewed in much the same way as rational dress – controversial, eccentric and in open conflict with 'ladylike' behaviour. As we have seen, that popular perception had quickly changed was a result of one key factor: it became fashionable. Evidence of how the norms of fashion could promote rational dress was clearly in evidence in nearby Paris, where bloomers were commonly seen as being chic and modern, meaning they were widely worn by female riders. 'The recent attempt to come back to the skirt for lady cyclists can be said to have failed miserably in Paris,' commented *Cycling* in 1896, as 'nearly every rider thinks the Rational dress the only suitable one for cycling, and it is everyday becoming more general'.[65]

Aiming to move Britain in a similar direction, members of the Rational Dress League and sister organisations implemented a variety of approaches. One was to emphasise the attractiveness of rational costume by wearing designs which were ornamental as well as practical. At a rational dress meeting in Reading, Lady Harberton was reported by a newspaper to have commented:

> People, and men particularly, had got it into their heads that 'rational' dress meant undecorated dress. She thought that the fault the advocates of 'rational' dress had made was that they did not make the dress sufficiently attractive. If they could manage to combine plenty of bright colours with 'rational' dress, she thought they would go a long way towards lowering opposition against it.[66]

Implementing the approach outlined by Harberton, the outfits worn by members of the organisations she was associated with were certainly far from being 'undecorated'. Newspaper articles on a widely publicised group ride from London to Oxford in September 1897, where the Rational Dress League and six other institutions aimed openly to promote more practical forms of female cycling clothing, provide wonderful descriptions of the types of apparel worn by members. While we must allow for editorial licence, a range of colourful outfits were clearly on show. The *Cheltenham Examiner* described

> ... a lady attired in a natty rational costume consisting of a short black jacket, enlivened with red lapels and collar, an open white shirt front, red necktie, voluminous bloomers, and a straw hat with a red ribbon.[67]

Another newspaper that covered the event reported:

> One costume was a creation in white, topped by a large white befeathered hat, a corded silk Eton jacket, and very full white bloomers, white shoes, and black silk stockings with white embroidery. There were visions in pale green, in pink, in fawn, in fact in all the colours of the rainbow, while one club of some half-dozen fair enthusiasts sported shepherd's plaid blouses over light grey knickerbockers, black stockings, and neat tan shoes, sailor hats and badges completing the uniform.[68]

Those involved in this 'gathering of the clans' also used the event as an opportunity to implement other promotional methods. While the involvement of Viscountess Harberton was a clear indication of upper-class support, it was recognised that this base needed to be expanded. Large numbers of distinguished and highly 'respectable' individuals had made cycling stylish in Battersea Park – the anticipation was they could work their magic again with rational dress. Invitations for the ride, which included an overnight stay at the Clarendon Hotel, were extended to:

> Lady Colin Campbell, Lady Randolph Churchill, Madam Sarah Grand, Lady Richardson, Lady Henry Somerset, Miss Eva Maclaren, Miss J. Harrison, The Countess of Warwick, Lord Coleridge, Colonel Saville, Mr W.T. Stead.[69]

Although not all those invited turned up, the staging of widely reported events such as this did achieve some noticeable successes. Following the ride a leading high-society paper, *The Queen*, produced an editorial in favour of rational dress:

> Looking on calmly and dispassionately, we find that the ordinary woman appears just as well in bloomers as in a skirt; and, as far as the usefulness is concerned, there is no comparison. We may try all kinds of patent skirts, but there is always danger attaching to them in a wind. It is merely a matter of time. Bicycling has come to stay, and bloomers will soon be accepted among us.[70]

Regrettably, this proved to be one of many false dawns. Aristocratic riders steered clear of bloomers, both unconvinced by attempts to make them more stylish and having little need for the practicality they offered when pedalling around inner-city parks. At the other end of the class spectrum, there was little let-up in the abuse and ridicule that poorer sections of society directed at those attired in rational dress. Despite their best efforts, the core group of supporters remained concentrated in the affluent middle-class groups led by Lady Harberton.

With the benefit of hindsight, it is hardly surprising that the Rational Dress League failed to spread its appeal downwards. Attempts to make rational dress a more glamorous costume inevitably pushed up the prices of the outfits members wore and promoted. The first edition of the *Rational Dress Gazette* advised its readers to purchase a tailored knickerbocker costume costing at least £3 10*s*.[71] This represented a price well beyond what many would have been prepared to pay for an actual machine, let alone the clothing they wore when out riding.

Despite the powerful image of a female cyclist in knickerbocker trousers, this evidently is not a straightforward tale of radical change and emancipation. After failing to take on with most female cyclists, rational dress had little chance of instigating the wider transformations in women's clothing that its supporters were hoping for. Furthermore, it would also be a mistake to view Harberton and her supporters as being modern-day feminists. Her description of the Surrey Hotel's bar parlour where she was forced to take lunch as a 'pigsty', and another prominent individual claiming she didn't 'care two pence for the great unwashed', reminds us that their outlook was informed by upset class sensibilities as well as a commitment to gender equality.[72]

This is not to underplay the exceptional bravery and resolve they showed in the face of such antagonism and adversity. Nevertheless, we should not force a story of sweeping advances led by a few individuals well before their time. A more realistic picture emerges when we focus on smaller but still significant developments in women's cycle clothing. While few openly wore knickerbockers, by the end of the 1890s it was increasingly common for these garments to be worn underneath divided skirts, which were split at the back and the front to provide freer movement to the legs.[73] This represented a significant advance on the outfits worn at the beginning of the decade, and also retained a sense of the risqué and adventurous. Gwen Raverat would later recall:

> We were then promoted to wearing baggy knickerbockers under our frocks, and over our white frilly drawers. We thought this horribly improper, but rather grand, and when a lady (whom I didn't like anyhow) asked me, privately, to lift up my frock so that she might see the strange garments underneath, I thought what a dirty mind she had. I only saw once a woman (not of course a *lady*) in real bloomers.[74]

The extent to which these developments supported wider efforts towards female emancipation is a matter for debate. The impact of bloomers was certainly limited − it was not until the First World War, when the land movement created a more pressing need for practical female clothing, that women's legs would again be seen in public spaces other than the seaside or the stage.[75] But at the same time, skirts continued to be developed and modified after the turn of the century so that they were better suited for housework, athletics, and other active pursuits.[76] Cycling's continued popularity among women depended on outfits which were far more practical than those worn by early female riders. And while not all who wore rationals identified themselves as 'New Women', the fact that Harberton would later raise her voice in favour of women's suffrage highlights the continuities which existed between two movements that challenged the denial to women of the freedoms and privileges enjoyed by men.

The struggle to develop more practical forms of women's cyclewear is hugely important to the tale of 1890s cycling, both in terms of allowing female riders to better enjoy the pleasures of riding, and for its wider links with female emancipation. But as we've touched upon at various points, the history of cycle clothing extends far beyond this. For all those who took to the pastime, a central concern was not just how well suited their outfits were for pedalling a bicycle. The ways in which it could be used to show off their affluence, taste and style were all also of significant value.

From a historical perspective, it can be difficult to know how best to appraise this. After all, women's greater freedoms and opportunities are much easier to celebrate than people looking to affirm their place within a restrictive social hierarchy. Was it a positive thing that the bicycle so strongly fed into people's preoccupations with how they were perceived by others?

A case can certainly be made for the end justifying the means. The huge increase in bicycle production stimulated by its newfound fashionable reputation later did much to drive prices down and create an accessible second-hand market.[77] Without it, members of the working classes would have been forced to wait much longer before they could afford the price of pneumatic-tyred safeties. As well as this, figures such as Sarah Grand, feminist writer and supporter of rational dress, saw plenty of positives in the activities of aristocratic riders in Battersea Park:

> When it is considered consistent with conventionality for the society girl to cycle, the way is made clear for the more practical lovers of the pastime to open up new pastures, and extend their devotion to the wheel to long country rides.[78]

Rather than getting bogged down in these discussions, however, it is best to bat away the question they seek to answer as irrelevant. Other than a few metropolitan hipsters, most of us now avoid the attire of our 1890s counterparts. At the same time, a concern with our clothing and appearances remains very much with us. The success of modern high-end brands such as *Rapha*, creators of the 'finest cycling clothing and accessories in the world', is in no small part a consequence of the status associated with their logo. The fact that bicycles and cycle clothing are consumer products means there will always be a ready audience for the marketing techniques developed by 1890s bicycle manufactures. It is easy to see modern parallels in a *Cycling* article which described a rider going out for the first time on his new machine:

> Observe the patrician scorn for every last year's machine that comes along, and the aha-a-hated-rival air when another equally new with his own runs past. Notice the agony when a plebeian spot of mud jumps up on to the enamel, the joy when someone praises the beauty, the wrath when another says, 'Thirty pounds, did you say? H'm, yes, *maker's weight* perhaps.'... You try it next time you meet a chap who tells you his safety only weighs so much. But I shouldn't do so if you value his friendship.[79]

In short, showing off is as fundamental to cycling as free-wheeling down sweeping hills, punctures, or mid-ride café stops. The elegance of bicycles as machines, the public nature of riding them and the opportunities this provides for displaying yourself to others, mean it is embedded in the very essence of the pastime. And while we can find ready amusement in the clothes worn by our nineteenth-century counterparts, it is easy to imagine that future generations, looking back on modern-day cycle outfits and the much-discussed phenomenon of middle-aged men in Lycra, will share a similar reaction.

But then again, there is far more to pedalling a bicycle than the opportunity to pose in public. After all, this is rarely the first emotion that accompanies mastery of a machine in childhood. Now able to propel ourselves forward faster than ever before, who can forget the thrilling sensation of speed and movement?

2

Racing, Superstars and Scorchers

There can be little dispute that in recent years, British cycle racing has undergone the most extraordinary development. From boasting the grand total of zero Tour de France winners in the 109 years of its history before 2012, Chris Froome and Bradley Wiggins now hold five yellow jerseys between them. Mark Cavendish is within sight of Eddy Merckx's record for the number of stage wins, and in 2011 became only the second British Road World Champion. Chris Hoy, Jason and Laura Kenny, Victoria Pendleton, Wiggins and numerous others have achieved phenomenal success in the velodrome, helping Britain win twice as many cycling gold medals than any other country in Rio 2016. Such has been the level of achievement, I fear that by publication the list above may already be outdated.

Given these recent triumphs and the relative lack of success which preceded them, it is easy to view British cycle racing as something of a modern phenomenon. To focus only on Tour de France wins and Olympic golds, however, would seriously misrepresent Britain's racing heritage. Even if the period when British cycling turned a corner is dated back to Chris Broadman's Olympic gold at Barcelona in 1992, this represents a minor time-span within the broader history of cycle racing in this and many other countries. After all, two hundred years ago young men had already started pushing themselves along in hotly contested races atop their 'dandy chargers'.

Such events were, of course, far removed from modern cycle racing. At first glance, perceptions of 'penny farthings' as quaint and old-world machines might make you think the same conclusion could be extended to high-wheeled bicycle contests. Such a notion, however, is quickly dispelled when one reads descriptions of cycle races and racing cycles from the 1870s and 1880s. Take, for instance, this passage by A.J. Wilson from a book published in 1887:

> For racing purposes, specially constructed bicycles and tricy-
> cles are used. These are identical in design with the ordinary
> roadsters, but they have no brakes to check the momentum,
> and their every part is made as light as possible, the thinnest
> gauge of steel tube being used; with very fine spokes; the
> lightest of bearings and forgings; and, in fact, with their every
> part reduced in weight as far as is consistent with the strength
> requisite to withstand the strains of racing on smooth and
> level paths.[1]

By the time Wilson was writing, cycle racing had already become a highly developed 'cutting edge' sport. Competition between manufacturers to produce the fastest, most advanced racing machines meant that a significant amount of technological skill and innovation was involved in their design and manufacture. Crowds flocked to witness the thrilling new marriage between an advanced piece of technology and seasoned athletes who were using it to engage in closely fought high-speed contests. As early as 1875, an estimated 15,000 spectators attended the Boxing Day 'championship' races at the Molyneux Grounds in Wolverhampton.

The excitement and provincial pride created by bicycle races is referred to in James Hawker's *A Victorian Poacher*:

> Riding at this Period was all the go… the People of Leicester nearly went mad with Excitement. We had in Leicester two Champions of the world on the Tall Machine… and for some time Leicester was Kept alive by the performances of these Cracks. They defeated all comers.[2]

Well before the invention of the safety bicycle and pneumatic tyre, cycle racing was firmly established as a spectator sport. But as with the bicycle's newly acquired fashionable status, the 1890s represented a boom period for racing. With cyclists now covering distances in previously unimaginable times, popular interest reached new heights. Specially built racetracks appeared across the country to host an ever-growing number of 'meets'. Cycle-based contests of strength, speed and stamina also fed off growing consumer demand for bicycles and cycle products.

Fashion and racing shared the essential quality of placing the bicycle firmly in the public eye. Although aristocratic bikists were certainly not known for exerting themselves awheel, their promenading around parks, like cycle races, drew large crowds of spectators and attracted widespread press coverage. The two very different activities imprinted a similar image of cycling in the popular imagination. By covering distances at increasing speeds record-breaking riders had, from the 1870s onwards, done much to create an image of the bicycle as an exciting new piece of technology. In shedding its lingering standing as a 'poor man's horse', the 'craze' of the mid-1890s further cemented its reputation as a highly desirable and 'modern' machine, placing it at the forefront of late nineteenth-century life.

The two branches of the pastime also enjoyed a more material relationship with each other. As we have already seen, the clothes worn by male road racing cyclists attracted criticism and comment, while rational dress was strongly associated with female racers, who were particularly in want of its practicality and freedom of movement. From the early days of cycle racing a profitable relationship had existed between racers and those involved in the cycling trade. Manufacturers realised that riders achieving success atop their brand of bicycle was an excellent means of advertising themselves to more everyday users, meaning that they were eager to sponsor and then promote the achievements of leading racers of the day.

Cycle racing and the consumer market which surrounded the bicycle were then united by a strong set of symbolic and practical ties. By coming together in a variety of different ways, they fostered much of the enthusiasm and anticipation, not to mention controversy, that surrounded cycling in this period. From speeding riders adding to the concerns of those worried about the 'respectable' public image of cycling, to female competitors offering an even more public challenge to traditional notions of 'femininity', racing added further fuel to the debates we explored in the previous chapter.

Finally, exploring the cycle race scene as it had developed by the 1890s not only paints a broader picture of the huge social impact of the bicycle in this period. The racing culture which exists in Britain today can be directly linked to customs and attitudes established during the late nineteenth century. Journeying back to this period allows us to understand why Britain's roads aren't home to 'classic' cycle races. It also contextualises the gulf in status and earnings which exists between male and female racers – and amidst the criticisms and controversies now threatening to engulf the sport – reveals that scandal has always been a part of British cycle racing.

Road racing and the 'scorcher'

On his cycle came the scorcher
Swiftly through the quiet street;
Well he knew the way to torture
Passers whom he chanced to meet:
Brushed the skirts of timid ladies,
Startled men, who stared aghast,
And consigned him unto Hades
As he shot, like lightning, past.[3]

When exploring the landscape of 1890s cycle racing in Britain, the best start-ing place is not the elite record-breakers at the top. Rather it is those who, for a variety of reasons, were seen to inhabit the bottom. Within the world of late nineteenth-century cycling, there was no figure viewed with more angst or aggression than the so-called 'scorcher'. This was a catch-all term used to describe cyclists who were seen to travel excessively fast when out riding in public, frequently bringing them into conflict with other road users.

Ill feeling towards 'scorchers' is evidenced throughout the 1890s press. The *Leeds Evening Express* quoted one well-known police officer who described a typical scorcher as being 'a straight-haired, thin-jawed, wild-eyed idiot, with his back humped like a mad tom-cat's tail, who bears down the road or street with no regard for the safety of others'.[4] Similarly evoc-ative imagery was used in a series of letters published in *The Times* in 1892 under the heading 'A Tyranny of the Road'. The correspondent whose original letter provoked several responses bemoaned the fact that he would no longer be able to enjoy his walks along the lanes of outer London if

... the gentlemen of the wheel are allowed to continue their present style of racing along at the rate of 20 miles or more an hour, not to speak of hill descents, when it is the practice for a number of them, spread across the road, to rush down at headlong speed, more like a horde of Apaches or Sioux Indians, conches shrieking and bells going; and woe betide the luckless man or aught else coming in their way.[5]

Although complaints such as these tended to border on the hysterical, it is not unsurprising that 'scorching' riders produced a strong public backlash. The invention of the safety bicycle and pneumatic tyre not only gave cyclists an ever-increasing presence on Britain's roads, they also created a very real increase in the speeds they could travel. At a time when most other road users were slow-moving pedestrians or horse-drawn vehicles, growing numbers of cyclists capable of moving at '20 miles or more an hour' quickly whipped up widespread fears and anxieties. Another *Times* correspondent stated:

> There was a time when the attitude of cyclists was unobtru-
> sive and even apologetic; this was when their number was
> comparatively small. Now, however, possibly in accordance
> with the proverb, 'L'union fait la force' [unity is strength],
> their recreation, the healthful advantage of which all are
> willing to allow, has degenerated into a veritable 'tyranny of
> the road'.[6]

Cycling's growing popularity did mean that criticisms such as these were grounded in a certain amount of reality. But then as now, complaints about cyclists could themselves easily degenerate into simplistic arguments that lazily grouped all who pedalled their machines into one reckless body. 'A great proportion of the non-rid-ing public seem to think that it is one of the pleasures of cycling to knock some old woman down then ride off,' complained a member of the Tottenham Cycling Club,[7] while a *Cycling* editorial, published in response to discussions in *The Times*, criticised attempts to caricature the average cyclist as

> ... a bloodthirsty desperado speeding on with dire and determined intent to slaughter men, women and children in his wild career. Of course we have all met the kind of garrulous cycle hater who would pen such spiteful words as these, and we will simply say 'Booh!' to it.[8]

When it came to the issue of 'scorching', however, the cycle press was far less sympathetic. As we saw in the previous chapter, speeding cyclists who abandoned 'the clothes and customs of a respectable individual' received severe criticism from fellow riders. Lacy Hillier painted the vivid picture of the 'would be scorcher' as one 'loose in garb and speech, given to horse play and Hooliganism, addicted to slang and sweaters'.[9] Hillier, along with most other middle-class authorities who were preoccupied with maintaining the 'respectability' of the pastime, had little time for those who brought cycling into public disrepute through their clothing or riding styles. Classist language was often present in criticisms of scorchers, with one Cyclists' Touring Club member stating:

> There are, I regret to say, far too many of those selfish individuals a-wheel who think the road is wholly and solely for their use, and it is such low bred cads... that we have to thank for the universal bad feeling which at present exists between the cyclist and the general public.[10]

For middle-class cyclists grappling with the 'bad feeling' directed towards cycling by the public at large, the easiest way to resolve this issue was to lay the blame squarely at the door of the lower classes. Falling prices of ordinaries and other older models of bicycle during the 1890s had resulted in increasing numbers of poorer riders entering the pastime, and they were soon dubbed the sport's 'rougher element' by the cycling press.[11] To characterise this group, cycle journalists often used the fictional figure of ''Arry'. 'Arry was a lower-class Cockney character invented by *Punch* editor E.J. Milliken, who related his various misdeeds to his friend Charlie in comic poems published in his paper. In ''Arry on Wheels', written in 1892 and inspired by the complaints being published in *The Times*, he wrote:

> But there's fun in it, Charlie, worked proper, you'd 'ardly
>   emagine 'ow much,
> If you ain't done a rush six a-breast, and skyfoozled some
>   dawdling old Dutch.
> Women don't like us Wheelers a mossel, espech'lly the
>   doddering old sort
> As go skeery at row and rumtowzle; but, scrunch it! that
>   makes 'arf the sport! ...
>
> My form is chin close on the 'andle, my 'at set well back on
>   my 'ed,
> And my spine fairly *'umped* to it, Charlie, and then carn't I
>   paint the town red?
> They call me 'The Camel' for that, *and* my stomach-capas'ty
>   for 'wet.'
> Well, my motter is hease afore helegance. As for the liquor,
>   – you bet![12]

There was, however, a major issue with making 'Arry solely responsible for damaging the 'respectability' of cycling. For such a perspective ignored the harm frequently done by his wealthier brethren several rungs up the social ladder. This was recognised in a letter by one *Cycling* reader:

> I cannot make a practical suggestion to improve ''Arry from the New Cut': I simply pass him with the remark, 'Language disgusting, manners none,' but it has been borne upon me very strongly of late that a radical defect exists in a proportion of the ordinary, middle-class, well-educated army of cyclists.

The writer went on to illustrate the point with the example of his friend 'Juggins'. In his day-to-day job in the city, Juggins was 'a very nice, affable fellow, of good attainments, well read… to all intents and purposes, a gentleman'. However,

> When Juggins, shortly after mid-day on Saturday, dons his cycling clothes, he undergoes the transformation so well illustrated in the case of 'Dr. Jekyll and Mr. Hyde.' By the time he has ridden five miles, and warmed to the exercise, the thin veneer of civilisation drops off. He harks back to some low ancestor – in a word he becomes little better than a savage.
>
> You, Mr. Editor, can meet Juggins on every main road leading out of London every week-end. He, and his club companions, 'spread-eagle' the road, and not only insult the pedestrian, but do not spare their fellow cyclist who happens to be in their way, although Juggins always rides on the wrong side. This is not an overdrawn picture.[13]

Such criticisms of 'savage' riders who spread-eagled the road were most strongly associated with the organised contests that took place on public roads. Road races, in which members competed to be the first to reach a pre-arranged point, had long been a feature of cycling club life. By the late 1880s, however, this had become an increasingly controversial subject, encompassing the same set of issues around respectability and the public image of cycling as individual 'scorchers'. Not only did races feature numerous competitors, but 'pacers' also supported contending riders by creating slipstreams they could ride in. As such, races frequently involved large groups of cyclists packed closely together speeding along Britain's roads. A *Cyclists' Touring Club Gazette* editorial in 1894 blamed road racing for 'nine-tenths of the odium' directed towards cyclists by the public:

> Like imps let loose from the lower regions, these pests swarm over our highways, transforming them into scenes of brutal butchery, and striking terror into the hearts of every peace-loving pedestrian.[14]

Not only was road racing dangerous, it was strictly speaking illegal. In this period, police were employed by local authorities to haul up cyclists engaged in 'furious riding', defined as 'riding and driving furiously, or so endangering the life or limbs of any person, or to the common danger of passengers in the thoroughfare'.[15] This was a law that could be easily applied to organised road races, and it was also the case that in many counties, particularly those close to London, such events were expressly forbidden.[16]

Faced with such opposition, the National Cyclists' Union (N.C.U.), the body responsible for overseeing cycle racing in Britain, distanced itself from road-based competition. In 1888 it ceased to provide official recognition for road records and directed its executive council 'to do its upmost to discourage road racing'. Throughout the 1890s the issue rumbled on until in 1897 the union banned any rider seeking a racing licence from participating in road races. The absence of N.C.U. support, combined with opposition from the police and public, meant the sport was effectively pushed underground.[17]

This is not to say it disappeared altogether. Several prominent clubs, supported by the Road Records Association, a breakaway group set up after the N.C.U. refused to recognise road racing, continued to host and promote races. One of the most significant developments in this regard came in 1894 when F.T. Bidlake and a group of other riders racing out of London sped past a trap, causing the horse to rear and the cyclists to crash into a ditch. The owner of the horse immediately raised the matter with the chief constable of Huntingdonshire, who, blaming the cyclists for the incident, swiftly banned cycle racing in the rest of the county.

Bidlake realised that it would not take much for similar accidents to occur in other counties, with bans subsequently imposed throughout the country. His response was to promote what was then a rarely used alternative: time trialling. As this race format did not involve groups of riders racing together, but rather setting off separately to see who could cover a pre-planned route fastest, it was a form of competition far less likely to result in accidents or attract police attention.

Despite these improvements, because time trialling involved cyclists 'furiously riding' along public highways it remained outlawed by the N.C.U. and could still land competitors in trouble with the police. Organisers were therefore required to think up innovative ways of ensuring these events attracted minimal public attention. No public-ity was given to them beforehand, with details of the date, time and place they were to start kept a closely guarded secret. The Bath Road

Club offered the cryptic advice to competitors for their 50 Miles Open Handicap that it would 'start at 2.45 p.m. *to the minute*... 300 yards north of the 36th Milestone from London, on the Great North Road, two miles north of Hitchin'.[18] Routes were revealed to riders only on the day of an event, which typically started early in the morning so as to avoid the police and other road users.

Significant efforts were made to ensure riders were as inconspicuous as possible. Competitors were required to wear dark clothing unlikely to attract public attention, while clubs such as the Anfield Bicycle Club specifically requested that members 'avoid all appearance of racing through towns'.[19] In the event of these tactics failing, riders had an excuse ready if they were pulled over, namely that they didn't know anything about an organised race, officer, but were merely out enjoying a vigorous early morning spin. To make this more believable, it was requested that they fit bells to their machines to give the impression of being casual, everyday cyclists: members of the Bath Road Club were instantly disqualified from races if a 'full-sized bell' was not present on their bicycle.[20]

Given the exhaustive planning involved in road racing, it is not surprising that those involved in the sport saw their efforts as something of a moral crusade. Celebrating outdoor racing for how it pitted riders against the elements, dodgy roads and the constant risk of machine breakdown, its supporters remained steadfast in their belief that 'road racing, despite the diatribes of the envious, the interested, and the incapable, remains undoubtedly the grandest and purest branch of the sport'.[21] Gazing out from the moral high ground, one member of the Anfield Bicycle Club philosophised:

> The spirit of competition will last as long as human nature
> is what it is, and the most determined anti-road-racer is

sometimes seen doing his best in an impromptu dust-up on the highway, and chuckling at having left some less speedy individual. Promiscuous scorching is far more annoying to the public than the well-arranged road race over a secluded course.[22]

Members of clubs such as the Anfield Bicycle Club do deserve a huge amount of credit for keeping the sport alive in Britain. But the tactics they were forced to use, which ensured road races took place on 'secluded courses' out of the public eye, meant the sport had little to no chance of taking off and gaining a wider following. These circumstances continued well into the twentieth century – it was not until 1966 that the Road Records Association allowed races to be publicised before they were held.[23]

By contrast, in France and other European countries road racing developed along an altogether different route. On quieter rural roads, 'scorching' cyclists avoided the same harsh criticisms directed at them in Britain. French roads were also administered by a central national body, not various local authorities. This meant they were in much better condition than British highways, and it was also far easier for race organisers to acquire the permission required to hold long-distance events.

French cycling authorities were also far more eager to exploit public interest in road races, and the money they could generate. France's cycling governing body, the Union Vélocipédique de France, extended open arms to bicycle manufacturers looking to sponsor races, while routes were planned to begin and end in major cities in order to maximise crowds and potential audiences. The marathon Bordeaux-Paris race, in which competitors raced the 580-kilometre route in a single day, quickly became a major national event, with French winners achieving hero-like status across the country. This period also saw the development of 'stage' races: events which took place over several days across a variety of routes. The Tour de France, first held in 1903, grew naturally out of the 'modern' forms of road racing which were nurtured and celebrated in France during the 1890s.[24]

On the two sides of the Channel, the closing years of the nineteenth century were a period when popular attitudes to road racing underwent a major divergence. The consequences of this are still very much with us to this day. While Britain has an established road racing culture, this remains centred around early morning time trials between talented amateur enthusiasts. Although recent years have seen contests with a more continental feel – such as the 'Tour de Yorkshire' – becoming increasingly popular, its title tells you all you need to know about the heritage of large-scale races on public roads in Britain. Unlike in France or other countries on the Continent, events such as these have never truly been a source of national pride, interest or celebration. In the absence of this, it is difficult to avoid a sense not only that little has changed, but that little ever will.

Taking to the path

While road racing slipped away from public view in Britain during the 1890s, the opposite was true of track cycling, a flourishing sport both at home and abroad. Tens of thousands of spectators were frequently recorded at races that attracted extensive media coverage. While meets had traditionally been held on temporary surfaces put down in football, cricket or athletic stadiums, as the 1890s progressed they increasingly took place on specially built tracks proliferating in Britain's towns and cities. With their surfaces layered with cement and asphalt, and raised

8 Competitors photographed before the start of the 1892 Cheadle Cycling Club Championships.

9 Racers Tom Linton and J.W. Stocks photographed in front of the grandstand before the start of a one-hour race at the Catford track (1896).

banks at either end so riders could 'slingshot' onto the home and back straights, these new venues enabled riders to push themselves and their machines to the absolute limit.

Stadiums not only gave seasoned competitors the opportunity to race on state-of-the-art tracks, they also catered for a large viewing public. London's Wood Green track, opened in 1895 after being financed by the businessmen Walter Gamage, was typical in including a covered grandstand on the home straight seating 1,500 spectators, a smaller 300-seat stand on the opposite side, and plentiful standing room which raised the overall capacity to 10,000. Further facilities included a restaurant and licensed bar, a café with outside seating, and outdoor spaces for traders to set up stalls. Catering for a variety of spectators, stadiums also offered a range of seating and pricing options. At the Catford race track in London, for instance, a reserved seat in the stand cost five shillings, two shillings and sixpence secured admittance into the unreserved enclosure, and admissions into the grounds alone set you back a shilling.[25]

Unsurprisingly, events that placed a wide cross-section of society in close proximity had a tendency to upset middle-class sensibilities. A *Cycling* editorial in 1895 pruriently protested that competitors not only 'race in the most scanty attire, bare arms, almost bare legs, the short breeches at no time reaching the knee… and so loosely fitted that the rider finishes as often as not with bare loins'. It also complained that after a race they would

> … stroll about in front of the grand-stand, and elsewhere, in this barbarous guise, sometimes, as we have ourselves heard, calling forth shouts of coarse chaff from the rougher portion of the crowd, to the embarrassment of the refined. To make matters, if possible, worse, the skins of these gentry,

so freely exposed, frequently exhibit unmistakable marks of long abstinence from the bath, and the general effect is thereby by no means improved.[26]

Crowds which contained both the 'rougher portion' and 'refined' elements of society reflected the growing numbers who had the time and money to enjoy commercial entertainment in Britain's towns and cities. During the late nineteenth century, workers began to benefit from Saturday half-holidays, while the Bank Holidays Act of 1871 designated four bank holidays in England, Wales and Ireland, and five in Scotland. Coupled with increasing real wages and a fall in the price of basic commodities, more people than ever before had the means to seek out fun, diversion and excitement in their leisure time.[27]

Like the theatre or the music hall, track cycle races were designed to cater for this growing demand. Typically held on Saturday afternoons or bank holidays, they provided a variety of entertainments. Military bands would be invited along to perform popular songs between races and help stoke up a carnival atmosphere, while stadium vendors catered for those wishing to eat, drink and make merry over the course of an afternoon. Other athletics events also featured alongside the cycle races, with novelty races such as handicapped contests between cyclists and runners used to offer lighter forms of entertainment.

But undoubtedly the real attraction of race meets were the contests which took place between the cyclists themselves. Marrying skilled athletes with an exciting new piece of technology, track races made for exhilarating athletic spectacles as competitors used their machines to test the boundaries of human-powered speed and endurance. Combined with the thrills and spills of closely-fought matches on specially banked tracks, cycle racing offered drama and edge-of-the-seat excitement to all those in attendance. The *Sheffield and Rotherham Independent* reported on a series of contests that took place at London's Olympia:

> It is a thrilling sight to see the riders fly round at the most appalling angle, with wheels overlapping, the outer riders actually overhanging the others. The whole spectacle has an unreal appearance. It is more like some strange picture. The men look so small, and flash along so noiselessly and rapidly, and there is such absolute stillness in the centre of the vast amphitheatre.[28]

At least, this is what was supposed to happen. Riders had in fact long since cottoned on to the fact that in any race that was not an all-out sprint, the last place you wanted to find yourself was racing out in front. In doing so you created a 'slipstream' for those behind to ride in, allowing them easily to match your speed while saving a considerable amount of energy. This meant that in the closing stages of a race they had plenty left in reserve to overtake you and cross the line first.

As all competitors wished to avoid this, races billed as battles between the fastest athletes on the planet could quickly descend into farce. This was vividly brought to life in an article in *The Saturday Review* covering a large international race meet that took place in Glasgow's Celtic Park in 1897:

> The Glasgow sky was at its clearest, so that when the sun shone out upon the vast amphitheatre − black with 30,000 spectators − in the midst of which was the gleaming red track and the green sward dotted with kilted bandsmen, the eye was abundantly satisfied and the mind stirred to expect great deeds.

In the short distance races which followed, however, these expectations were 'curiously falsified'. For upon hearing the starting pistol the competitors did not speed out of the blocks, but rather

> ... crawled off with slow, wobbling, hesitating movement, each man watching his neighbour with an alert, sidelong glance. Sometimes, indeed, they seemed to stop altogether, and always they crept along, climbing the steep sides of the track to lengthen the journey, and playing a game which would lead a novice to believe that the last man across the tape was to be the winner.

For crowds (and indeed journalists) not clued up on the science behind slipstreaming, such a performance provoked a less than favourable reaction. It was not until the final lap that

> ... the racers laid themselves down to the work and pedalled furiously, skirmished and shuffled and bored to get position on the inside edge of the track, and came flying round to the winning-point at the speed of an express train. They had covered a mile in something over six minutes! Whereupon the disgusted spectators declared that an active man could almost have walked the distance in that time.[29]

The *Derby Mercury* was equally disparaging in its coverage of the event, stating how six minutes was 'about the time that would be made by a very indifferent amateur, who had begun cycling late in life and was trying hard to ride slowly'.[30] In the early years of the 1890s, a period when 'loafing' riders were particularly prominent at race meets, the issue was one which vexed cycling authorities. Numerous proposals were put forward both to make races more in line with Victorian sporting ideals of 'playing up and playing the game', and to increase their attractiveness to the wider public.

Most successful of these was an approach which followed the example of road races and made use of 'pacers'. This prevented races from turning into 'disgraceful crawls', as each competitor cycled behind a designated machine that provided them with a slipstream. Racers were typically paced by friends who might belong to the same cycling club, or specially trained pacing teams. These often rode specially built 'multi-cycles' that accommodated several riders, which increased both their speed and the size of the slipstream enjoyed by the cyclist pedalling behind them. Leading manufacturers footed the considerable bill which came with training and maintaining teams of 'pacers', using them to ensure success for the riders whom they sponsored. The ways in which pacing had transformed the nature of cycle racing was recognised in an article in *The Hub* in 1896:

> The pacing of the racing cyclist is at the present day not only a veritable science, but an extensively followed profession. Hundreds of men are earning their living as pacemakers; and the exhibitions of speed and skill given week by week on our faster tracks prove to what a high pitch of perfection the art has now been carried... In most paced races, the riders go at absolutely top speed all through.[31]

Growing understanding of the science behind pacing, along with developments in track design, ensured records were being continually broken and pushed forward. By the turn of the century, gasoline and electrically powered pacers, capable of travelling far faster than teams on multi-cycles, were becoming increasingly prevalent on race-tracks.

In 1899 the British rider Albert Walters covered over 1,000 kilometres in the 'Bol d'Or' twenty-four-hour race, averaging speeds of over forty-two kilometres per hour racing behind a gasoline-driven tandem.[32] Record-breaking feats such as these, achieved by riders travelling at great speeds over mammoth distances, ensured cycle racing was a sport which continued to enjoy widespread media coverage and remained firmly in the public eye.[33]

As with any good spectator sport, however, interest in track cycle racing was not just stimulated by record-breaking feats and triumphs. Crowds flocked to races not only to be blown away by the achievements of individual riders, but to watch, cheer and support the riders *themselves*. In a period when cycle racers were achieving widespread renown and recognition, their names proved a massive draw for partisan audiences. The birth-place, racing style, temperament and achievements of competitors became widely known and created the wider narratives and story-lines that were critical to ensuring the sport's mass appeal.

Who then were the individuals at the centre of late nineteenth-century cycle racing? To understand these characters and the following they enjoyed among spectators, the first attribute in need of consideration is one not actually listed above. For it was this trait which did far more than any other to determine the reception they enjoyed on the start line. Before the question of ability came a far more important issue: their gender.

'The Brighton female scorcher'

Whatever their social background, very few late nineteenth-century women enjoyed a childhood as full of sporting opportunity as Tessie Reynolds. Born into a working-class neighbourhood in Brighton, she had, along with her ten siblings, learnt from a young age how to cycle, fence and box under the tutelage of her father, R.J. Reynolds, who worked as a bicycle dealer. Well acquainted with people working in the cycling trade, Reynolds also grew up close to the Preston Park Velodrome as well as the popular London to Brighton road record route, meaning she was regularly exposed to the feats of record-breaking riders.

In September 1893, aged just sixteen, she resolved to follow in their footsteps by cycling the 106 miles to London and back in under ten hours, thus setting a female record for the route. Attired in rational dress and paced by a group of male cyclists who belonged to her father's cycling club, she began her journey at 5 a.m., setting off from Brighton Aquarium.[34]

From a sporting point of view Reynolds's ride was certainly a triumph. By quarter past nine she had already reached Hyde Park Corner and departed back home to Brighton. After three short stoppages during the return leg she returned to Brighton Aquarium at 1.38 p.m., meaning she had broken the previous female record with a time of just over eight and a half hours. In the days that followed her ride, Tessie found herself at the centre of widespread media attention and coverage. Very few journalists or correspondents, however, found anything complimentary to say about the time she had achieved. Dedicating a highly opinionated editorial piece to her ride, *Cycling* thundered:

> Every cyclist who truly loves the sport, every lady rider who has striven, in the face of many difficulties, to spread the gospel of the wheel amongst her sisters, every wheelman who has managed to retain a belief in the innate modesty and sense of becomingness in the opposite sex, will hear with real pain, not unmixed with disgust, of what it would be moderate

to call a lamentable incident, that took place on the Brighton road early last Sunday.[35]

The debates which swirled around Reynolds were, inevitably, partly a consequence of clothing she wore. At the time of her ride female riders were still a rare sight on Britain's roads, with many questioning 'whether ladies can with propriety ride the bicycle'. The image of Reynolds on a diamond-framed 'man's' machine wearing knicker-bocker trousers therefore caused widespread consternation. *Cycling* described her outfit as being 'of a most unnecessary masculine nature and scantiness', which 'no woman possessed of the instincts of a true lady would care to appear in public with, let alone ride through London streets'.[36]

The real issue, however, was that Reynolds had been widely seen racing along public roads. When men 'scorched' in public this was bad enough, but for a woman to engage in such an activity was to push against an even more formidable boundary. For *Cycling*'s editorial team, members of the 'fairer sex' engaging in strenuous physical activity was in no way compatible with their 'innate modesty and sense of becomingness'. This was a mindset shared by many other middle-class publications. 'The woman who allows herself to be seen hot and red with exertion, and panting from want of breath, loses much of her feminine dignity,' argued *Hearth and Home* in 1895, while a writer in another article stated:

I sincerely hope that an era of feminine record-breakers is not upon us, for if racing either on road or track is to be the order of the day, then goodbye to the popularity of cycling, for nothing is more calculated to bring that pleasant pastime into disrepute.[37]

'Feminine recordbreakers' were thus another aspect of cycle racing that middle-class authorities found too controversial to stomach. Even organisations such as The Lady Cyclists' Association, active supporters of rational dress, had the stated objective 'to discountenance cycle racing amongst women' so as not to 'lower the tone' of women's cycling.[38] The National Cyclists' Union also offered firm opposition, refusing to award official licences to female cyclists wishing to compete in track events. Writing in 1896, one of its vice-presidents, Edward Beadon Turner, echoed well-worn arguments:

> Organised public races for female riders must have a bad effect upon the development and extension of quiet, healthful road-riding and country touring for women and girls... the spectacle of a crowd of women frantically contending for supremacy on bicycles is not an edifying one, the remarks and criticisms of the sort of crowd which patronise such an exhibition are still less choice, and the combination of these two factors must go far to influence against the innovation of feminine cycling.[39]

Up to this point, then, the story of female cycle racing has noticeable parallels with that of men's road racing. Both found themselves condemned by Britain's largest racing body and were subject to fierce criticism from the cycling press and sections of the public. But it is at this moment that their two paths diverge. For rather than being pushed underground, track races between female competitors were, by the mid-years of the 1890s, attracting significant amounts of public interest. Covering the final day of a series of women's races which had taken place at London's Royal Aquarium, a large indoor venue known for hosting music hall and variety acts, *Lloyd's Weekly Newspaper* reported:

10 Photograph of Tessie Reynolds wearing rational dress and riding a men's racing bicycle (1893).

11 Group of female professional racers including Hélène Dutrieu (second from left) photographed alongside their trainer Choppy Warburton at the Royal Aquarium (date unspecified).

The contest, which is the first of its kind held in England, proved of a decidedly popular character, and so large were the crowds that flocked to it each evening that the managers of the building greatly increased the prices of admission to the reserved portions of the building; in fact, for last evening some of the seats were sold at a guinea and half-a-guinea each.[40]

Popular enthusiasm for women's racing is reflected in the careers enjoyed by leading female riders. As well as competing on tracks across Britain's major towns and cities, they also travelled over to France where there was a firmly established racing scene. By winning cash prizes and receiving sponsorship and salaries from leading manufacturers, female racers could enjoy a considerable income. In an interview with *The Hub*, Rose Blackburn, an established English rider, described how in eight weeks during 1895 she had ridden 3,049 miles and taken £140 in prizes and salary, a figure which equates to around £8,500 in today's money.[41] The potential earnings of female racers are further reflected in another *Hub* article, which reported that a group of professionals who travelled over to Britain from France in 1896 had made more during their stay than their male counterparts.[42]

Given the opposition expressed against women's cycle racing, we need to ask how the sport managed to establish itself. To begin with we must recognise that not all shared in the reproaches put forward by middle-class authorities. Following *Cycling*'s editorial on Reynolds's ride, numerous correspondents wrote in offering support and praise for her achievement. 'I fail to see what harm a ladies' race, or a ride from Brighton to London and back in 8½ hours, can do to the cause of ladies' cycling,' wrote one female reader, while John Keen, a famous ex-racing cyclist, stated:

I have nothing but admiration for her pluck and splendid consti-
tution which enabled her to place on record such an excellent
performance, and for her courage in daring to appear in public
in a new, novel, and at the same time, sensible costume.[43]

Although similar viewpoints continued to be aired in the years
following Reynolds's ride, growing interest in women's racing was,
regrettably, not a result of Keen's admiration becoming widely
established. Women who raced in public wearing rational dress
continued to defy convention and receive widespread criticism and
reproach. They remained, in short, something of a curiosity. And
it was for precisely this reason that they excited so much popular
interest.[44]

While all forms of track cycling benefited from the unusual sight
of speeding riders pedalling around steeply built tracks, women's
racing was a truly unique branch of the sport. Because so few women
had ridden ordinaries, there was little tradition of female cycle
racing in Britain. Race organisers, unconcerned about the 'respectability' of cycling and keen to vary the entertainment provided to
their audiences, soon realised the potential for including ladies' races
within their programme of events. Josiah Ritchie, director of the
Royal Aquarium in London, told *The Hub* that 'as men's races were
no novelty, I put on the ladies' races first, and they were an instantaneous triumph, although at the start the trade and the Press were
dead against us'.[45]

The novelty of female cycle racing was not just a result of the
sport's lack of history or heritage. For a public unaccustomed to the
sight of women engaging in competitive sporting contests, matches between female riders were especially original and at odds with
convention. Newspaper reports spared no detail in reporting on

the falls, crashes and collisions which took place in women's races, while *Bicycle News* offered the opinion that 'probably three-fourths of the audience at the Royal Aquarium "ladies' cycle races" attend in the hope of seeing what one man, the other day, termed "a holy smash".'[46]

As well as this, races afforded voyeuristic male spectators the rare opportunity to witness women sitting astride a saddle while wearing knickerbocker trousers and other revealing costumes. Drawing on promotional efforts first developed in France, organisers ensured that competitors wore outfits such as breeches and short skirts, which invariably flew up the leg over the course of a race. Writing in 1893, one *Cycling* columnist, seemingly immune to the baser instincts of his fellow man, offered this opinion:

> Lady-racing is 'not *nice*'. France has given us a useful object-lesson in this respect... who wants to see Herne Hill sprinkled in the evenings with panting, perspiring young females in very light attire, riding all over the track, turning corners on the top of the banking and coolly stopping now and then to adjust a too tightly-laced corset, which is the spectacle frequently enjoyed of late, by the lookers-on at the Vélodrome Buffalo?[47]

Reading such descriptions, it is easy to view the track races between female riders as risqué forms of popular entertainment rather than serious sporting contests. The fact that they often took place in between theatrical and music hall acts in popular venues such as the Royal Aquarium does little to dispel such a viewpoint. Writing in 1895, the *Birmingham Daily Post* argued:

> The competition which commenced on Monday last at the Westminster Aquarium [another name for the Royal Aquarium] is to be welcomed in so far as it determines the character of racing by women, placing it upon a spectacular and professional basis, thus tending further to check any attempt to introduce more ordinary forms of competition upon recognised race tracks. As a race the affair cannot be taken seriously.[48]

Many would have agreed with the writer's assessment of the race, and women's competitive cycling more generally, as a form of popular exhibition. It can be questioned, however, how many of those who critiqued these events would have fared against any of the competitors. For few could match the speed or stamina of professional female riders, who away from the race track lived the life not of racy public entertainers, but serious full-time sportswomen.[49]

Their sponsorship agreements ensured they received specialist coaching and were required to undergo constant training to keep themselves in peak condition for races. Interviews with leading riders also reveal the careful attention they gave to their diet, typically steering clear of alcohol other than the occasional 'glass of champagne after a big ride'. Their dedication to the sport is reflected in their race performances: Rosa Blackburn averaged speeds close to twenty miles an hour when breaking the 100-mile female record in Birmingham's Bingley Hall in 1896.[50] Hers was a time soon bettered by the Scottish racer Clara Grace, who enjoyed phenomenal success in this period. Take, for instance, her race performances as recorded at the end of 1895:

23rd December... 1st, National 100 miles Championship

24th December... 1st, 20 miles, Westminster

26th December... 1st, 25 miles (starting from scratch in
handicapped race)

27th December... 2nd, 50 miles (starting from scratch in
handicapped race)

28th December... 1st, 40 miles (starting from scratch in
handicapped race)[51]

As well as setting numerous records on the race track, Grace enjoyed notable triumphs on the road. A few years after Reynolds's return ride from Brighton to London she did the same route in 7 hours 14 minutes, taking over an hour off her record. Grace's time was then improved by Maggie Foster, who in 1897 clocked 6 hours 45 minutes and continued to break women's time trialling records into the early 1900s (by this point *Cycling* had given up condemning such rides in favour of quietly ignoring them).

No British racing cyclist from this period, however, enjoyed a career quite like that of the Belgian rider Hélène Dutrieu. Described as 'a pretty, pale, dainty little girl plucky to the point of foolishness', she initially found fame through her performances on the track. Training alongside her two brothers who were both well-known professional sprinters, she broke the women's hour record in 1893, while also winning early versions of the Women's World Championship in 1896 and 1897. As time progressed, however, she left behind competitive cycling in favour of variety stage performances in which, nicknamed the 'Human Arrow', she showcased daredevil tricks and stunts on her bicycle.[52] One of her most famous acts, 'Flying the Flume', was described by a newspaper:

Mounted on a bicycle, the girl slides down a steep shoot, some seventy feet long, like a flash, and then up a short incline. Then there is a gap of 40 feet wide in the track. So terrific is their momentum, however, that bicycle and girl take one blood-curdling leap over this 40 foot of space, alight on the track on the other side, and then dash into a rope stretched across the path. When Mademoiselle does this feat there is only one unconcerned person, and that is herself.[53]

After going on to perform the stunt with a motor cycle, and then an automobile, Dutrieu's attention turned towards the Wright brothers' newly invented aircraft. Her small build, extensive cycling experience and death-defying bravery made her the perfect candidate to fly what were highly unsafe and unreliable machines. With no prior training, she became only the fourth woman in the world to acquire an official pilot's licence, going on to set world records for altitude reached, distanced covered and time spent in flight.

Although she caused a minor scandal when it was revealed that she did not wear a corset while flying, Dutrieu's gender was of little hindrance to her flying career. As well as performing in races and exhibitions all over the world, she became the first winner of the Coupe Femina (Women's Cup) for long-distance flights and was awarded the French Legion of Honour in 1913. Now nicknamed 'the girl sparrow hawk', she offered the opinion that flying would turn out to be 'as common among women as it is becoming among men', before warning potential suitors that 'the man who wants me must catch me in the air'.[54]

Clearly the path Dutrieu forged after leaving cycling would not have been possible without her strength of character or phenomenal daring. However, her career trajectory is telling of the attitudes that surrounded female racing during this period. For it was only after she abandoned the sport in favour of one where women's participation was subject to far less conservative opposition that Dutrieu rose to such fame and prominence. While competitors and their sponsors undoubtedly took women's racing seriously, interest in individual racers was rooted not in personalities or achievements, but rather in the novelty of what they represented. When this wore off in the early 1900s the sport underwent a rapid decline and quickly went out of fashion.

This is not in any way to detract from the dedication and physical abilities of leading female competitors, or the phenomenal courage they showed in participating in a sport subject to so much hostility and lewd comment. As Claire Simpson has recognised, their short-lived popularity represents one of the few times in history when women have earned more than men operating within the same sporting code.[55] Nevertheless, the cultural norms that informed competitive track cycling in this period made it inevitable that its leading lights were all exclusively male. Those at the pinnacle of men's racing achieved levels of fame, income and acknowledgement that leading female riders could only dream of.

As with road racing in Britain, the repercussions of this are still very much with us today. Modern female riders do of course enjoy far greater professional status than their 1890s counterparts. But it is one thing to say things have come a long way from what is now well over a hundred years ago; it is quite another to assert a fundamental break from the past. In the absence of a well-established race scene – the first women's Olympic cycling event was not held until 1984 – competitive female cycling is a sport still playing catch-up with its male equivalent. Its race calendar remains far smaller, which means strong disparities continue to exist in earnings and career opportunities. While efforts have been made to bring down walls created more than a hundred years ago, one has only to read recent assertions by Olympic gold medallist Nicole Cooke that British Cycling is overseeing a sport 'run by men for men' to gain a sense of the barriers that still remain.

Zimmy and the Boy Wonder

While it can be difficult to find anything 'modern' in late nineteenth-century female cycle racing, when we examine the world of male racers a different picture quickly emerges. We've already seen that in this period track cycling became a widely reported and popular form of spectator sport. It therefore comes as little surprise that the careers that leading male riders enjoyed contained many elements of the present-day sporting 'star'. Looking back this time to the 1893 racing season, as described in the Indianapolis *Wheelmen's Gazette*:

> The racing cracks have thriven wonderfully the past season.
> They have been the idols of the land, and the whole cycling
> world has bent knee to do them homage. Rival cycling clubs
> have fallen over themselves in their frantic efforts to secure
> the attendance of the stars.[56]

The hero status enjoyed by successful 'cracks' often extended beyond their homeland. With cycle racing attracting more and more popular interest in Europe, North America and Australia, they increasingly became global figures whose names and achievements spread across continents. The international celebrity enjoyed by male cycle racers is nowhere more apparent than in the careers of two hugely successful riders: James 'Jimmy' Michael from Wales, and the unforgettably named American, Arthur Augustus Zimmerman.

Away from the race track, you could hardly have found two more different figures. Michael's physique was defined by a childhood growth disorder which meant he was little over five feet tall, earning him nicknames which ranged from 'Boy Wonder' to the less flattering 'Little Jimmy' and 'Midget Michael'. Zimmerman, by contrast, was

12 Early photograph of 'the boy wonder', Michael, and his trainer (date unspecified).

13 Zimmerman (on left) photographed alongside George Lacy Hillier at
the Herne Hill Track (1892). As may be inferred, the pair did not enjoy
an easy relationship.

long and lanky, his slim six-foot frame resting on long, powerful legs. Differences in build were mirrored in personality. Michael was well known for being confrontational and prickly, a character who regularly came into conflict with others and effectively abandoned his wife to pursue his racing career. Conversely, 'Zimmy' enjoyed a happy home life and was an affable, if often shy and retiring, figure. Major Taylor, who would go on to become one of America's first major black sports stars a few years later, described how Zimmerman, upon setting a new one-mile world record,

> ... received an ovation which would have caused most any man to have swelled visibly, but he paid no attention to it, and ambled along to his dressing room like some green, country boy who never seems to know where to put his hands... When he is called upon to make a speech or respond to an address of presentation, he blushes and stammers and gets as nervous as a school girl, and yet he can go out on the track and whip the world riding.[57]

The two men also came from very different social backgrounds. Son of a wealthy New Jersey real-estate owner, Zimmerman was set for a career in business and the law until his obvious athletic ability was spotted and nurtured by his brother-in-law, Joseph McDermott. A county champion jumper, talented at both the long and the high jump, Zimmerman slowly turned his talents towards high-wheeled bicycle racing in the late 1880s.

Michael's route into cycling was, by comparison, rather less glamorous. Raised in the Welsh mining town of Aberdare by his paternal grandmother, he was at a young age put to work as a delivery boy for her butcher's shop. The large, heavy bicycle he was required to pedal, and membership of Aberdare's cycling club, provided the ideal preparation for his career as a racer. This small mining community was, remarkably, already home to Arthur Linton, an internationally renowned racer who had bought his first penny farthing from tips earned working as a haulier in Aberdare's coal mine. Arthur, whose brothers Tom and Samuel were also highly talented riders, mentored Michael and supported him both on and off the track.

Given their respective backgrounds, it is not surprising the pair chose two very different routes after making it in the sport. For most of his career Zimmerman raced as an amateur, defined by the National Cyclists' Union as 'one who has never engaged in, nor assisted in, nor taught any athletic exercise for money or other remuneration'. Having grown up as a grocer's assistant, Michael was far more eager to exploit the financial rewards that could be found in competitive cycling. He turned professional at the age of nineteen, allowing him both to race for cash prizes and to seek sponsorship from leading manufacturers.

But for all their differences, 'Zimmy' and 'the Boy Wonder' were united by their capacity to, in the words of Major Taylor, 'whip the world riding'. Michael came to prominence when, aged just eighteen, he appeared in the prestigious 'Surrey Hundred' at London's Herne Hill track. Up against 'the cream of the present day cracks at the distance', the little-known Welsh rider caused a sensation. As *Cycling* put it:

> No matter what the pace, no matter who thought they would like to sprint, Michael was always there, riding as easily as if on a club run. Who was this youth everybody asked, looking barely 14 or 15 years, who dared to hang on to London's speediest riders?[58]

At forty-six miles Michael, chewing on what would become his trademark toothpick, had already lapped the entire field. His time at the fifty-mile mark, just over two hours, broke the previous record for this distance. As he approached 100 miles, 'cheer upon cheer greeted his victorious parading around the track' as he continued to leave the entire field trailing in his wake. Finishing two miles ahead of his nearest competitor, Michael's time of 4 hours, 19 minutes and 39 seconds was a full ten minutes faster than the previous 100-mile record set by his mentor and fellow Aberdarean Arthur Linton.

Michael's unexpected triumph made him an overnight star, and provided the springboard for a highly successful career as a middle-distance rider. After turning professional later in 1894, he spent much of the next two years racing in France and continental Europe where there was a well-established professional racing scene. Making an ever greater name for himself, Michael triumphed in the 1895 world championships in Cologne, becoming the first professional world 100 kilometre champion. His youth and diminutive stature delighted crowds awed by the strength and staying power packed into his tiny frame. To quote German rider Walther Rutt:

> I saw Jimmy Michael win that first Championship at Cologne, when he was only 18 [Michael was actually 19]. I was 12 years old at the time. I remember telling my father when I returned home from the races that a 'boy' had won the world's paced championship! He was 18 years of age on that day and looked like a little boy.[59]

Following more European success, Michael found himself drawn by promises of the riches which could be found as a racing cyclist in America, and crossed the Atlantic late in 1896. Here he continued to break records and draw mammoth crowds eager to witness 'Midget Michael' take on the best America and the rest of the world had to offer. *Spalding's Bicycle Guide for 1898* described him as

> ... the most marvellous athlete the world has ever seen, for with his diminutive size he combines a power and ability that is gigantic, and during the last season has duplicated in this country his record in England, France and Germany. He has been the bright particular star of the match racing season. He has met defeat only once during the entire season, and he has met all who were brave enough to face him in a race.[60]

In just a few years, Michael had moved from grocer's assistant to an international sporting star. The fame and recognition he achieved is testament to how, in the closing years of the nineteenth century, there was no sport so firmly established on a worldwide scale as cycle racing. Admittedly, in Britain it struggled to draw quite the same crowds or generate interest on the scale of football, cricket or rugby. But it is inconceivable that if Michael had possessed the same talents in any other sport, he would have been such a global success. Only track cycle racing held the prospect of finding fame and fortune in Britain, continental Europe, America and Australia. And nowhere was this more apparent than in the career Zimmerman had enjoyed only a couple of years previously.

As with all celebrated sportsmen, Zimmerman's fame was built upon his extraordinary athletic ability. He was renowned for his elegant, upright racing style and ability to emerge like a rocket from the back of the field to take victory in the closing stages of a race. His coach and brother-in-law, Joseph McDermott, told the *New York Times*:

Zimmerman has one big advantage. He can stay in a race and go fast without pumping himself out, and then when the final effort comes he is faster than anybody. No matter how fast a rider finishes, if my boy is in good condition he will come ahead of him... Zimmerman is not a rider, but one of the wonders of the age.[61]

Although in his efforts to promote his rider McDermott may have ventured into hyperbole, 'Zimmy' was most certainly a phenomenon. He won over a thousand races during a hugely successful career, with his two most significant victories being in the ten-kilometre and one-mile sprints at the 1893 amateur world championships in Chicago. Befitting his widespread billing as 'Champion of the World', Zimmerman also achieved numerous triumphs racing abroad. In an 1892 tour of Britain, he won both the fifty-mile amateur national championship in London, as well as the one- and five-mile events which were held at Leeds. Looking back on all the American riders who had made their mark on British shores during this period, *Cycling* would later comment:

Of these the outstanding figure was the 'Jersey Skeeter', Zimmerman, who took the country by storm. He was the finest rider ever seen on the Safety from the other side, and will long be remembered by all who saw him. Compared with the English riders he had a graceful and easy style, sitting almost upright on his machine, and this made him appear to win from doubled-up men with ridiculous ease.[62]

After undertaking two more highly successful visits to Europe in 1893 and 1894, Zimmerman's career culminated in a tour of Australia in October 1895. The receptions he received at each city he travelled are testament to just how brightly his international star shone in this period. The *Australian Cyclist* reported his arrival by train into Adelaide:

> There must have been fully four thousand persons present to welcome the Jersey 'skeeter.' As soon as he approached the crowd he was cheered again and again, and he responded by raising his hat and bowing to the sea of faces.[63]

Similarly, several thousand people congregated near the train station to meet him as he entered Melbourne, before the city's mayor 'pledged the visitor's health with champagne'. Although Zimmerman struggled with illness throughout his time in Australia, causing him to pull out of several races, appearances he did make were received ecstatically. An estimated 30,000 spectators were reported to have witnessed his appearance at Sydney Cricket Ground, where they 'waved their arms in wild excitement at the phenomenal American Wheelman'. As in Europe and America, his athletic abilities and unassuming personality helped Zimmerman win widespread adulation. According to the Sydney *Referee*:

> Apart from being perhaps the most genuine champion America has sent us, Zimmerman is the best specimen of a man, in the proper sense of the term... he is cosmopolitan, as befits a world champion. There is no nonsense about him, no blow, no nothing in fact that is objectionable, and he wears his laurels with becoming modesty.[64]

If he had wanted to, Zimmerman could have made a fortune from his now firmly established international celebrity. Upon returning to America he was not short of tempting offers: one promoter offered him $500 a week to ride in Paris, another $10,000 for a three-year contract. But after much speculation, and despite the fact he was only in his mid-twenties, Zimmerman walked away from cycle racing early in 1896. This was most probably a result of burn out after years of continuous travel and racing on the international racing circuit, reflected in his steadily declining performances from 1894 onwards. Although he made a couple of short-lived returns in 1897 and 1899, these amounted to little more than a few 'exhibition' matches against fellow 'old-timers' whose best racing days were well behind them.[65]

Jimmy Michael retired from competitive cycling at an even younger age. Disillusioned with the track racing scene, he took the unexpected step of abandoning the sport to seek out a career in horseracing. This was a decision Michael would live to regret, as he achieved only a fraction of the success he had gained as a racing cyclist. Although his small size and background in racing should have stood him in good stead as a jockey, his tendency to refer to reins as handles and stirrups as pedals is ample evidence that he was far more at home on a bicycle than a racehorse.[66]

While Zimmerman enjoyed a comfortable retirement as a hotel owner in New Jersey, Michael's life had a much more tragic ending. After paying a heavy financial price for his horse racing career, he returned to cycling in the late 1890s. Replicating past triumphs proved to be difficult, however. Having struck up a friendship with the French rider Jean Gougoltz, who struggled with alcohol addiction, Michael increasingly turned to drink himself. In 1904, after an evening of heavy drinking on board a steamship from France to America, Michael was found unconscious in his cabin. Aged just twenty-nine, he was declared dead by the ship's doctor. The cause, a brain convulsion, was thought to be the result of a head injury he had suffered racing in Berlin a couple of years previously, which had been exacerbated by his drinking problems.

Zimmerman's and Michael's careers do not just tell the story of the prestige and opportunities that were open to racing cyclists. Their early retirements, and Michael's struggles later in life, provide a powerful reminder of the flip side of international stardom. For all its financial rewards, the mental and physical demands which came with a nomadic lifestyle built around pressured and gruelling athletic contests inevitably exerted a heavy toll. They were two of the first individuals to realise the price one paid for being idolised as a 'champion of the world'.

It was not just superstar cyclists, however, who struggled to cope in the rapidly expanding world of competitive track cycling. The celebrity, wealth and international fame of Zimmerman and Michael symbolised a racing scene far removed from the high-wheeled contests that had taken place only a decade previously. For Britain's cycling authorities, who had governed the sport since its early years, this posed a challenge which they increasingly struggled to deal with. Without official support, road racing had been effectively pushed underground. Refusing licences to female riders had by and large ensured women's races remained removed from established race tracks. But exercising control over a sport now awash with money and commercial and public interest was always going to be a far tougher proposition.

Amateurism and professionalism

In most modern sports, 'amateurism' is a concept which has long since gone out of fashion. While leading professionals may continually frustrate us with 'amateurish' performances, the wealth and huge incomes they often enjoy could not contrast more strikingly with the other definition of 'amateur': someone who engages in sport on an unpaid basis.

That we are now generally unfamiliar with this meaning is not surprising: in most sports its heyday was in the later years of the nineteenth century. During this period, amateurism was not an outdated concept, but rather an ideal that exerted a powerful force on all forms of organised sport in Britain. Understanding its appeal, both to those who oversaw and to those who participated in cycle racing, is therefore key for completing our picture of the sport in this period.

As a sporting philosophy, amateurism was born in England's public school system. From the mid-nineteenth century onwards, schools such as Eton, Rugby and Harrow placed an increasing importance on their young charges participating in games such as cricket and rugby. 'The battle of Waterloo was won on the playing fields of Eton' became a widespread maxim, as sports were celebrated for the physical and moral development they imparted to young men. Their importance ultimately lay not in victory or defeat, but in fostering self-reliance, discipline, and a sense of 'fair play' in those who would one day need to 'govern others and to control themselves'.[67] Writing in 1891, R.J. Mecredy applied this ethos to cycle racing when he asked rhetorically:

> What better training for the race of life could there be for a youth? All the same qualities he will require for success, and the chivalry and friendliness towards other competitors, and the power to take a beating like a sportsman, without excuse or murmur, with no feeling of discouragement, but rather a determination of doing better next time, will all stand to him in his struggle with the world, and will not only help him to be successful, but will ensure him the respect and confidence of those with whom he is brought into contact.[68]

Away from the public school or university, amateurism played a critical role in carrying these ideals into the wider world. Earning a salary from a pastime or competing for cash prizes was seen to detract from the 'higher' purposes of sport, inevitably leading to cheating or 'foul play' by encouraging competitors to focus too heavily on beating their opponents. A true amateur was, in the words of Richard Holt, one who 'should not seek any advantage over an opponent that he would not expect his opponent to take over him'.[69] Corinthian Casuals, a football team established in 1882 and consisting of England's best public school footballers, embodied this principle to the full. Not only did they instantly remove one of their players if an opponent went off injured, but they also refused to score from penalty kicks on the principle that no player would ever deliberately foul an opponent. Captain C.B. Fry proclaimed it to be

> ... a standing insult to sportsmen to have to play under a rule which assumes that players intend to kick, hack and push opponents and to behave like cads of the most unscrupulous kidney.[70]

Such lofty idealism would undoubtedly have struck a chord with the Bicycle Union (later the National Cyclists' Union), the body set up to institutionalise and govern cycle racing in 1878. As with the Football Association or Rugby Football Association, this was an organisation created by public school men eager to place amateurism at the heart of their sport. Its first chairman, Gerard Cobb, was still president of the Cambridge University Bicycle Club when he took on the role, while fellow club member Ion Keith-Falconer, 9th Earl of Kintore, acted as chair of its racing committee. One of the union's first actions was to create a clear set of criteria to differentiate the amateur from the professional rider. After much debate, it settled on this:

A Professional Bicyclist is one who has ridden a bicycle in public for money, or who has engaged, taught, or assisted in bicycling or any other athletic exercise for money.[71]

It must be noted that by contemporary standards, the Bicycle Union's classification was controversially forward-thinking – the Amateur Athletic Club, which had previously governed cycle contests, explicitly excluded anyone working as a 'mechanic, artisan or labourer' from competing as an amateur. This definition had created a farcical situation at the first official amateur championship race in 1871 where seventeen of the twenty entries were barred from competing, ensuring the race was a walkover for public school H.P. Whiting.[72]

Although the Bicycle Union's definition was less exclusive, it was still a means by which its middle- and upper-class members sought to place themselves at the head of the newly established sport. While the union stated it was open to professionals as well as amateurs, in practice it was heavily biased towards amateurism. It largely ignored the working-class professional racing scene developing in the Midlands and north of England, instead focussing on promoting and regulating amateur races between clubmen centred in London and the south.[73]

Bicycle clubs, whose members had been heavily involved in the founding of the Bicycle Union, played an integral role in organising race meets between amateur riders, who typically belonged to and rode in the name of their institution. During a racing season that lasted from early spring until late autumn, clubs hosted races on athletics tracks which they would hire out for the day or, as time progressed, in purpose-built stadiums that they ran and managed. To tackle the inherent tension in charging spectators to witness amateur contests, proceeds from these events were then donated to charities or invested back into club activities.

As with making expensive cavalry-style uniforms mandatory on weekly runs, upholding the ideal of amateurism was a means by which these institutions differentiated themselves from the 'great unattached' and allowed members to affirm their wealth and class status. Indeed, for the young men who raced as amateurs, it can appear that their primary motivation was not to live out ideals of fair play, but rather gain the social kudos which came with publicly refusing to race for money. For while this revenue stream was cut off, riders were quick to exploit more discreet means of making an amateur racing career a financially rewarding one.

In the absence of cash prizes for races, rewards such as silver cups, plates and vases were often on offer to race winners and runners-up. Upon acquiring these trophies, there was little to stop an enterprising rider from cashing them in for their monetary value. Furthermore, manufacturers were also keen to sponsor successful amateurs. In early forms of now widely used arrangements, cycle makers provided well-known racers with travel expenses and state-of-the-art racing equipment, which allowed them both to showcase their machines at race meets, and to associate their brand with the leading riders of the day.

Well before the 1890s these factors had made 'shamateurism' a widely recognised issue that caused frequent headaches to both the Bicycle Union and its later incarnation as the National Cyclists' Union. 'Makers' amateurs', a term used to label those who received sponsorship from manufacturers, were a far cry from the gentlemanly Ion Keith-Falconer, who while being 'the best bicyclist in England' was described as one who treated cycling primarily as an 'amusement', delighted in 'feats of strength and endurance for their own sake' and carried his honours with a 'charming modesty'.[74] Decrying the 'farcical amateur law' in 1891, *Cycling* said it was widely acknowledged that

... the bulk of the men who race on road or path are the reverse of millionaires, and that any system of amateurism must be saturated with humbug. Everyone knows the glaring cases of makers' amateurism – prominent riders are daily pointed out, who are subsidised by makers, and so the farce continues from year to year.[75]

As cycle racing continued to grow in popularity and became an increasingly lucrative business, the 'farcical amateur law' became increasingly absurd. This is typified in the situation Zimmerman found himself in at the end of the 1892 season. During this time he had become the proud owner of 'twenty-nine bicycles, several horses and carriages, half a dozen pianos, a house and a lot, household furniture of all descriptions, and enough silver plates, medals and jewellery to stock a jewellery store'. The following year yielded another impressive haul, as he was reported to have won fifteen bicycles, fifteen jewellery rings, fifteen diamonds, fourteen medals, two cups, seven studs, eight watches, a tract of land, six clocks, four scarf pins, nine pieces of silverware, two bronzes, two wagons and a piano, whose combined value was estimated at $15,000.[76]

Zimmerman was also fully exploiting the earnings that could be made as a 'maker's amateur'. In the rapidly growing bicycle market manufacturers were now investing large sums in sponsoring leading riders, and they did not come any more successful or highly regarded than 'Zimmy'. His numerous achievements and positive public image enabled him to sign lucrative deals with Raleigh, Dunlop, Brooks and other well-known brands, which ran large-scale advertising campaigns openly associating themselves with his success.[77]

Such arrangements were in glaring contradiction to Zimmerman's status as one who did not 'engage, teach or assist' in cycling for money.

These incongruities meant that in 1893 the National Cyclists' Union refused to recognise him as an amateur and barred him from competing as such in Britain. The following year he dropped the façade of amateurism altogether and registered himself as a professional. Given the ridiculous prize-giving system he had to endure, and his obvious connections with wealthy manufacturers, this was the natural route to take. With owners of race tracks across the world offering Zimmerman vast sums of money to compete at their venues, becoming professional enabled him to enjoy openly and more fully the financial rewards increasingly on offer to well-renowned racing cyclists.

Zimmerman's change in status reflected a racing scene that, by the mid-1890s, was progressively moving towards professionalism. In a now fully developed 'modern' sport, amateurism was coming to be seen in the same outmoded terms as it is today. As well as being at odds with the vast sums of money involved in cycle racing, there were obvious flaws in the idea that a mass spectator sport could be organised by amateur clubs whose unpaid committees worked in other full-time occupations.[78]

Yet it was this vision of cycle racing which remained the primary focus of the National Cyclists' Union. As an organisation, it remained committed to older ideals of amateurism, and a belief that 'true' amateurs could still exist in a sport awash with cash and commercial interest. Nowhere were these views expressed with greater conviction than by the most prominent of its leading members. In a sport full of strongminded individuals, they did not come any more assertive, dogmatic or opinionated than a character we have already encountered: George Lacy Hillier.

George Lacy Hillier

If there was an aspect of British cycling during the 1880s and 1890s which Lacy Hillier did not influence, or at least comment upon, then it probably wasn't worth knowing about. Pictured on page 78, he authored and contributed to numerous books on cycling, acting as a self-styled oracle for the pastime. In constant need of an outlet for his thoughts and opinions, Hillier put his journalistic flair to further good use as editor of *Wheeling*, one of Britain's most widely read cycle-based papers.

English high-wheeled amateur champion for 1881, he maintained an especially fierce interest in cycle racing. Further to serving as a committee member, Hillier acted as official judge and time keeper for N.C.U.-sanctioned races, was director of the Herne Hill track in south London, and created the London County Cycling and Athletic Club to arrange and oversee race meets. He exerted a powerful influence across the sport, and in 1887 his character was summed up by the *Irish Cycling News* in the following terms:

> Always heard before he is seen, Mr. H. discourses on every topic, and lays down the law in tones rather suggestive of bombast; he speaks to no one in particular but at everybody in general. [He] is a man of good physique; a fine, athletic, well-knit figure, topped by a face of rather Mephistophelian cast. The mustachios have an aspect of soldatesque ferocity; they are waxed furiously, and stick out in a bold, defiant way, as becomes an integral portion of the amateur champion of 1881.

But despite these 'peculiarities', the writer of the piece remained of the opinion that

> Mr. Hillier has much to recommend him. He is almost an
> insanely enthusiastic sportsman; he believes himself to be the
> guardian angel of English cycling; he has done a good deal
> for it, and he ever strives according to his lights, to purify
> racing and lift it above suspicion.[79]

Hillier was certainly a man who lived up to his reputation for bombastic idealism. Editorials in *Wheeling* claimed wooden race track surfaces were superior to cement or asphalt; an even less successful prophecy was that pneumatic tyres would soon go out of fashion as they represented no improvement on their solid predecessors. He also caused minor diplomatic episodes by calling into question the times set on French and American race tracks, arguing that for N.C.U.-recognised English records

> … every precaution is taken to ensure accuracy; and I prefer to
> accept these records and to stand by them rather than to accept
> records accomplished upon tracks of dubious measurement,
> clocked by excitable, not to say Chauvinistic timekeepers with
> watches of doubtful accuracy.[80]

Of the almost unending list of topics that Hiller conversed on, amateurism was the issue on which he most fully lived out his role as 'guardian angel of English cycling'. He viewed amateurism as nothing short of a moral crusade, and was steadfast in his views on the dangers of professionalism. In a piece published in *Cycling* in 1893, a time when growing numbers of professional riders were appearing on English race tracks, he argued:

> Speaking as one who knows, as one who has seen this sort of
> thing before, I am candidly of opinion that the whole thing

will slowly, but surely go the old road – that the divisible prize will be divided – that A. who wins by arrangement to-day will let B. beat him to-morrow – that the unlucky C. will be 'given a chance', lest he should lose his berth, and that eventually the whole class will drop down to the level of other professionalism.[81]

For all of Hillier's apparent cocksureness, it is easy to see why an upsurge in professionalism would have unnerved him. After all, support for amateurism from middle- and upper-class authorities had never stemmed solely from moral principles. In excluding less wealthy professional racers, it was a system which had also allowed them to exercise effective control over the sport. An established professional racing scene – drawing large crowds, attracting the best riders and supported by wealthy manufacturers – would inevitably challenge the N.C.U.'s position as overseer of competitive cycling in Britain. Hillier's approval of professional riders was therefore dependent on them having minimal involvement with the cycle trade, and taking up their 'natural and proper position' by accepting 'the patronage and countenance of the National Cyclists' Union and its constituents'.[82]

With a governing body focussed on maintaining an amateur system integral to its long-term survival, professional racing was left to flounder in Britain. Writing in 1900, a time when professional racing had 'run itself out' in Britain, what was now *Cycling and Moting* offered the opinion that 'if the ruling bodies had tried to help instead of hinder the pro, things may have turned out a little better'.[83] Throughout the 1890s, Jimmy Michael and other leading British professionals spent most of their careers racing away from home in France and America, countries which were much more comfortable with professionalism and offered far greater rewards to successful riders.

But while the N.C.U. saw off the threat of an established profession-al race scene developing in Britain, they could do little to reform the amateur system which remained 'saturated with humbug'. Successful amateurs continued to enjoy patronage from leading manufacturers, and although authorities made examples of a few prominent cases such as Zimmerman, they were powerless to stop the numerous others who gained financially from their racing careers. Another piece in *Cycling and Moting* from 1900 went so far to state that 'for all practical purposes, in the eye of the general public there is now no distinction between the amateur and professional racer'.[84]

The fact that Hillier and others in the N.C.U. remained unwavering in their commitment to amateurism, a system which was quite evidently completely unworkable, means it is easy to view them as self-interested and hypocritical. These were men who fiercely proclaimed the virtues of a sporting philosophy, all the while quietly ignoring the power, priv-ilege and – most controversially – wealth it gave them. Working as a stockbroker, Hillier made plenty of money from companies involved in the 'boom' of the 1890s, while his London Country Cycling and Athletic Club was a limited company that profited from organising amateur race meets.[85] In a fiercely critical article, the French cycling journalist Paul Hamelle asked:

What else is the theory of amateurism but a system of preju-dice, born of error, supported by lies… Blessed by the N.C.U., Hillier has made amateurism his credo, an article of faith outside of which there is only shame and misery for the obsti-nate sinner… From the heights of his intolerant dogmatism, he goes on passing opinions on people and events, making judgements which the facts hasten to contradict, without questioning his serene belief in his infallibility.[86]

At the same time, we should be wary before exercising too heavy a judgement on Hillier and the N.C.U. However much they benefited personally from the amateur race scene, they did possess a genuine belief in the more high-minded principles behind it. It is important to remember that they operated in a time long before the main purpose of governing bodies was ensuring Olympic success. In the absence of this worldview, their outlook was centred around a philosophy that, for all its classism, did focus on what most of us would still agree are the real purposes of sport: the pleasure and personal development it affords us. That they remained wedded to an amateur system that failed to place these values at the heart of competitive cycling was a consequence of naïve optimism, as well as their own powerful self-interests.

Although we can recognise the broader context that shaped Hillier's views on amateurism, it cannot be disputed that the outlook shared by himself and others at the N.C.U. was profoundly backward-looking. While we may acknowledge the principles which stood behind their support for amateurism, in remaining committed to this ideal they refused to engage with a rapidly changing, money-orientated race scene. Their most defining quality was, therefore, their conservativism. Further on in his article on Hillier, Paul Hamelle described him as:

> A high priest without any believers, who expends his substantial power blowing on dead ashes, a naïve supporter of lost causes who is inevitably opposed to any novelty… barring the road to progress and crying out to anyone whom he sees from far off, 'Halt! Who goes here? No entry!; Progress will pass through nevertheless. G.L.H. will at least have had the consolation of having sometimes stood in the way!'[87]

Until very recently, it would have been easy to rejoice in the dispari- ties between Lacy Hillier and those now at the top of British cycling. His backward-looking mindset appeared completely at odds with their widely publicised embrace of innovation and forward-thinking. Utilising sports science, data analysis and countless other means of gaining an edge over their rivals, they symbolised the rewards that could be gained by journeying down, rather than barring the 'road to progress'.

Furthermore, alongside Team Sky's highly professional exterior was a strong commitment to clean racing and restoring the reputa- tion of what had become a widely discredited sport. Pioneering, principled and hugely successful, both Sky and British Cycling were understandable sources of national pride and celebration.

Unfortunately, at the time of writing these reputations appear to have been irreversibly lost. Allegations of bullying and sexism within British Cycling have revealed much darker sides to their win at all costs mentality. And with questions about doping and illegal use of substances within Team Sky growing ever louder, what once looked like one of Britain's all-time great sporting stories has been muddied into some very murky shades of grey.

These developments can even encourage a re-assessment of Lacy Hillier and the principles which he so forcibly expressed. Whilst it may be proven that Sky have broken no rules in pushing therapeutic use exemptions for leading riders, they do appear to have stretched them right to their very limit. Although Lacy Hillier's basic outlook is well outdated, one feels that there was plenty of room for his high-minded idealism amongst those now overseeing the sport.

At the same time, it is worth remembering just how miserably he and the N.C.U. failed to create a race scene which embodied their amateur values. As we have seen throughout this chapter, Britain's modern race scene is still tied to the one that developed in the late nineteenth century. Amidst current allegations and scandal, this broader perspective leads one to wonder how much room such a demanding, intensively competitive and lucrative sport has for lofty ideals.

3

Club Life, Fellowship and Elderly Young Men

A much-used justification for studying history is that it allows us better to understand the present. By analysing the past and the forces which influenced its future direction, we may hope to gain a better understanding of the world we live in today. While this is surely true, sometimes the most telling insights come from people, events or institutions so wonderfully absurd that they can initially seem completely at odds with such a sensible rationale. And there are few better examples of this than nineteenth-century cycling clubs.

What we discover is a profound divide between the external facades of such clubs and the behaviour that club mates enjoyed once out of the public eye. From Flora Thompson's self-important 'townsmen out for a lark' who played leapfrog on the village green, to the keenly 'respectable' amateur riders who 'scorched' along public roads, we have already seen how their members' decorous outward personas could swiftly transform once in more secluded settings.

It is therefore not so surprising that the more familiar you become with club goings-on, the more evident, and indeed entertaining, such contrasts become. The Bristol Bicycle and Tricycle Club (B.T.C.), for instance, was one of many institutions from this period which took a keen interest in raising money for local and national charities, while also enjoying a close relationship with leading civic dignitaries. The Mayor and Dean of Bristol both attended and gave speeches at the club's end of year annual dinner in 1896, while famed solicitor Hugh Holmes Gore Esquire served as club president. Writing in 1897, he offered this opinion:

> Sport is the birth right of good workmen, not of idlers… all true sport goes in the direction of bettering us… [teaching] strength of limb, soundness of mind, fleetness of foot, accuracy of the eye, readiness in emergency, persistence in accomplishing an end.[1]

When it came to riding their bicycles, however, members wasted little time in leaving such lofty ideals at home. Only a month before Holmes Gore's assertion, they had cycled back to their native Bristol from neighbouring Gloucester. 'Persistence in accomplishing an end' was hardly the order of the day, judging by their multiple visits to pubs and inns on the journey home. Neither was 'accuracy of the eye', as evidenced by one member experiencing a nasty collision with a cow. Judging by his club mates' 'hilarious spirits on their return', much merriment was found in both these events which were, one assumes, not entirely unrelated.[2]

First-hand accounts such as this, which so brazenly contrast with the club's public reputation, are far from common when studying the past. The source material which we use to build up a picture of earlier periods has typically been passed through various filters to remove information that groups or individuals wanted to avoid sharing publicly. It is certainly hard to imagine the circumstances in which *Cycling*'s editorial team, keen to promote a respectable image of their pastime, would have featured such a riotous article within the pages of their magazine.

Fortunately for us, not to mention the writer of the piece, there was an alternative publication whose editor was more than happy to publish his account. Like many other institutions from this period, the Bristol Bicycle and Tricycle Club went to great lengths to produce a monthly gazette, designed to keep the wider membership informed of club happenings. Limited in its distribution and written by members for members, it was the perfect space for detailing what really went on during their rides. As we shall see later, these gazettes provide a quite remarkable set of historical records, giving honest, revealing, and above all else highly readable accounts of the adventures that 1890s cyclists experienced during their weekend excursions.

This rich source of information could alone justify a chapter on club caperings and goings-on. The significance of these associations, however, extends well beyond tell-all accounts of what went on out of the public eye. As institutions, the clubs also played a major role in a much broader set of social developments. From the all-male organisations which had defined early cycling, by the mid-1890s clubs had become another site where women's entry into the pastime was creating controversy and debate. During this period they also found themselves at the heart of Britain's growing socialist movement, as thousands came together to live out ideals of 'good fellowship' and use the bicycle to spread their message to new audiences.

With many clubs entangled with these radical new ideas, not to mention their wider efforts to court public recognition and approval, they played a key role in helping the bicycle acquire its modern, fashionable and controversial reputation. But possessing memberships who were eager to find enjoyment and gratification on weekend rides, their activities bring to life an aspect of the pastime which has so far been underexplored. That is, the pleasures and liberation which cycling brought to late nineteenth-century men and women.

Alongside the revelry enjoyed by Bristol B.T.C. men on their ride home, there are numerous other accounts revealing the weird and wonderful ways that members enjoyed their weekend excursions. Continuing the theme of enjoying liquid refreshment while out on runs, an account of 'one of ours' appeared in the gazette of the London-based Stanley Cycling Club:

> Tooling out on his steed one morning a few Sundays ago, he proceeded, like the busy bee, to cull here and there the good things which bestrewed his path. Lovely bunches of flowers, choice fruits, and – er – sodersandbrandies [sodas and brandies] he plucked galore, and thus bedecked, he wheeled round and started for home, with thoughts of the brightness and fragrance his lovely impedimenta would impart to aforesaid home.

Unfortunately, things then took a turn for the worse:

> But alas! and alack! As the poet sings, 'the best laid schemes of men and mice 'gang aft agley', and in this case the floral and fructiferous burden went very much 'agley', and our hero was last seen postured in the middle of the road with the spoils of war around his recumbent form.[3]

But while stories of ill-fated solo trips make for highly enjoyable reading, it is the relationships that members formed through cycling that are most fascinating. Why is it that from the bicycle's earliest years, so many have sought out fellow enthusiasts for their rides?

'As fine a body of young fellows as the eye could hope to witness'

There can only be one starting point for exploring this question: the early bicycling clubs whose activities so conspicuously took place on ordinaries. As well as establishing practices and traditions that would shape later club life, the distinctive nature of high-wheeled riding ensured that there has never again been a period when organised associations have been such a prominent feature of the pastime.

First and foremost, given the various risks which came with pedalling an ordinary, it was quite natural for young enthusiasts to band together for mutual support and protection. When heading out on journeys along poorly maintained and unfamiliar roads containing horse drivers and other 'roughs' known for their hostility towards cyclists, all the while perched atop a rickety machine, there was a most definite logic to the idea of safety in numbers. Looking back on this period, a member of the Bristol B.T.C. known as 'The Unchained' (of whose writings we will encounter plenty more later) reflected:

> The cycler of those days risked his neck, with the other portions of his anatomy, on what, in appearance at least, was a pair of cart wheels connected by a pump handle. On this fearful contrivance he ventured forth into the country, at that time, practically a *terra incognita* [unknown land] to the ordinary town resident. Like unto the Ishmaelite was he in the sense that every man's hand was against him.
>
> The street loafer and gamin considered him fair sport to chuck 'arf a brick at, and delighted to push a stick through his wheel, 'an honest peasantry their country's

pride' putting a clod of earth or a hayrake to the same noble use... As for consideration by other road users, why the very coster's or market woman's donkey considered it *infra dig* [beneath their dignity] to give an inch of road to the man on the wheel. No wonder, cyclists early found it necessary for self protection to go for their afternoon spins in company, and started bicycle clubs.[4]

A keen awareness of the dangers that attended pedalling one's 'fearful contrivance' into unknown and potentially hostile districts can be seen in the rules that governed the excursions of early institutions. Clubmen were required to travel in single or double file behind a recognised Captain, who distinguished himself from his fellow members by wearing a prominent gold badge. Orders to mount, dismount, or 'go ahead' were typically given by a 'bugler', who blew recognised signals on his instrument, while members also used whistles to communicate with each other. An early history of the Pickwick Bicycle Club, published in 1905, reflected:

On looking through the old minutes, one cannot help being amused at the kind of military discipline and method which ruled the excursions of the Club in those times, when to disobey the command of the Captain was a crime almost deserving of capital punishment; and when strict obedience was exacted to the riding rules and regulations laid down. In this connection, it appears that, in September, 1870, a system of signals by whistle was established, for the guidance of members when riding together on dark winter evenings. One sharp whistle signified '*caution*'; one long whistle, '*gathering or closing up*'; two whistles, '*stoppage*'; and three, '*distress*'.[5]

Further examples of the 'military discipline and method' which governed club life in this period can be found in the handbook of the Canterbury Cycle Club:

- That in all Club runs the Captain shall lead, and no-one shall be allowed to pass him without his permission.
- That the Captain shall have entire control during all runs, and (for the safety of the public) the power of compelling Members to slacken speed or dismount when passing horses &c.
- The Sub-Captain in all runs shall keep to the rear and look after stragglers, but in the absence of the Captain, shall take his place, nominating his own Sub-Captain for the time being.
- Any member desiring to fall out must report himself to the Captain or Sub-Captain.[6]

Other than for those unfortunate individuals who found it necessary to 'fall out', enforcing such stringent regulations made certain that members carried out runs as one united body. While guaranteeing mutual support, however, these rules offered little additional protection from the dangers of high-wheeled riding. 'Headers' and other accidents were part and parcel of pedalling an ordinary, whether or not you were riding in close ranks behind a Captain. Equally, as the historian of the Pickwick Bicycle Club observed, 'if a poor youth in distress were being assaulted by a rough, before he could blow three times he might get the whistle knocked down his throat'.[7]

It is fair to say that self-preservation was not the only factor influencing clubs' decisions to implement such stringent regulations. As with their cavalry-inspired uniforms, they tied into something which was equally dear to the hearts of young ordinary riders: showing themselves off. Adhering to military-style rules was yet another means by which clubmen sought to distinguish themselves from the 'great unattached' and conspicuously affirm their respectability and social standing to the wider public. The spectacle of groups of uniform-clad cyclists' high up on their ordinaries, pedalling in orderly lines behind their Captain and following the orders of a bugler was, at the very least, quite impossible to miss.

This was especially true of an annual event which did more than any other to bring cycling into the popular consciousness in this period. First held in 1874, the concept behind the 'Hampton Court Meet' was a simple one. After coming together outside the palace's famous Lion Gate, clubmen from various institutions would, at the sound of an 'advance' signalled by a bugler, proceed on a five-mile journey around Hampton's neighbouring roads and parks. An arranged procession close to one of London's most recognisable landmarks, the meet was designed to take cycling out of the rural settings in which it usually occurred and bring it firmly to the attention of city-based spectators.

Although the first event attracted only fifty participants, nearly 500 attended the following year as it quickly snowballed in popularity; by 1881 a reported 2,500 cyclists were taking part in a procession which featured 160 different clubs.[8] The impression this 'immense number of grey-clad phantoms' made on those in attendance is made apparent in newspaper accounts. A piece in *The Standard* described the start of the 1878 meet:

> Shortly after five o'clock – when the spectators could not have counted certainly less than 30,000, and might in all probability have amounted to double that number – the procession moved off to a note of Mr. Coppin's bugle, caught up and

repeated by other marshals along the line. The procession was headed by the dark blue Pickwickians, for the very good reason that theirs was the oldest amongst the clubs… Boldly they rode and well, these Pickwickians, as indeed did the great majority of the members of those other clubs mounted on their steel steeds.[9]

Moving on to illustrate the spectacle achieved by 'as fine a body of young fellows as the eye could wish to witness', the writer recounted, 'a really imposing and picturesque appearance they presented as two and two they floated past in their various uniforms, ranging in colour from sober grey and sombre brown to bright silver and orange or rich navy blue'. Voices of those who took part in the meets were of a similarly positive nature – the only cause of complaint tended to be the 'spirit of insolent independence' shown by members of the 'great unattached', who were grudgingly given permission to bring up the parade's rear.[10]

It is easy to see the attraction of club life, which offered both status and security, among young ordinary riders. There was, however, another feature of these institutions which exerted an even greater pull on prospective recruits. For those looking to find pleasure and gratification after long working weeks, there was little contest between the choice of heading out for their weekend excursions by themselves or in company. After all, few things are more likely to make such outings enjoyable than the presence of like-minded 'good fellows'.

14 Print showing the Pickwick Bicycle Club leading the Hampton Court Meet around the Diana Fountain in Bushy Park (1877).

15 Photograph of thirteen 'Pickwickians' with their bicycles, uniforms and club boaters (1885).

'Harmless young savages'

Even by the distinctive standards of high-wheeled clubs, few could claim quite the same level of individuality as the Pickwick Bicycle Club from London. Now widely viewed as the world's oldest active cycling club, from its earliest years it has held a prestigious place within the sphere of British cycling. As we saw from the piece in *The Standard*, the club's claim to be the oldest institution at the Hampton Court meets ensured 'Pickwickians' were given 'place of honour' at the front of the procession, proudly displaying their dark blue uniforms and ivory club buttons to the watching crowds.

The club's memorable name stems from the circumstances which surrounded its foundation in the summer of 1870. As told by the club historian, this was a time when 'Charles Dickens had but recently quitted for ever the sphere of his immortal labours' and in their 'sorrowful enthusiasm' for the great writer, members hoped to associate themselves with his name and works. Not only did they name their club after his first publication, *The Pickwick Papers*, it was further decreed that each member should take on a sobriquet taken from a character in the book, with the club captain to always be known as 'Samuel Pickwick Esq.'. Subsequently, every weekend in summer months he headed a motley crew which contained, among others, Mr Jingle, Tracy Tupman, Serjeant Buzfuz and Joe the Fat Boy (who was in fact one of the club's most successful racers) as they journeyed out of London and into the local countryside.[11]

The pseudonyms taken on by 'Pickwickians' speak volumes for the widespread popularity of high-wheeled clubs. As with countless other voluntary associations before and since, clubs were a breeding ground for nicknames, shared jokes and, above all else, sociability. It was difficult for clubmen not to bond and feel a strong sense of fellowship with those who shared in the incidents and difficulties that inevitably arose during weekend runs. Looking back on this period of cycling, one *Cycling* contributor recounted:

> It was wonderful what deeds of self-sacrifice were often performed; on some of the runs slight mishaps occurred, but the men were always eager to aid the distressed one, even to the extent of losing a pleasant evening, perhaps walking a considerable number of miles, and all in the innocent unconsciousness of performing an act of self-abnegation.[12]

The flip side of overcoming misfortune, of course, was sharing in the pleasures of finishing work for the week and hurtling through the countryside on your high-wheeled machine. At the beginning of his article, the writer in *Cycling* recalled:

> … those good old days when, all kindred spirits, we mounted our wheels and sallied forth, a jolly crew, revelling in our health, freedom, vigour, forgetting, for the time being, our worries or cares, if we possessed any, and living happily in the present. How, when we got out into the country, we became thorough, though harmless, young savages, laughingly meeting minor accidents to selves or machines with what was downright philosophy, and scorning the petty conventionalities of every-day life.[13]

For young men eager to release pent-up energy, club life came with many other opportunities to let rip. Away from cycling, a common form of entertainment they enjoyed were so-called 'smoking concerts'. These typically took place in a hired room in a pub or hotel, where clubmen were invited to spend the evening smoking and drinking to their hearts' content, as they each took turns singing popular songs of the day to a tune provided by an accompanying pianist. As one member of the London-based Argus Bicycle Club put it, 'the chief ingredients of a good smoking concert are plenty of smoke, high-class singing, jovial companions and sufficient liquids to combine these together in one fellowship'.[14] It could well have been after witnessing such an event that the journalist Paul Creston commented that although 'nominally a recreative and instructive evening of music', cycling club 'smokers' would instead degenerate

> ... into little better than a drinking bout; and the spectacle
> of numbers of young men sitting the long evening through
> in a reeking and tobacco-laden atmosphere, while they fill
> themselves with liquor to the accompaniment of vulgar and
> often obscene songs, is certainly the reverse of edifying.[15]

That drinking played no small role in facilitating friendship between clubmen can be seen in the large numbers who, once a year, spent a long weekend camping just outside Harrogate in the north of England. First started in 1877, the 'Harrogate Camp' served many of the same ends as the 'Hampton Court Meet', acting as a focal point for the coming together of members from different organisations. While Hampton Court was used to garner maximum publicity for club activities, however, Harrogate served a very different purpose. With each club setting up a spacious and richly adorned tent from which they played host to other institutions, the camp was renowned for its late nights full of smoking, singing and heavy drinking. Infused with an 'air of genuine bohemianism', those attending were frequently found attired in fancy dress, with a Stanley Cycling Club member detailing how at one camp he thought he had witnessed famous music hall singer Marie Lloyd

> ... dressed in such a bewitching costume, and occasionally blushing in such a becoming manner, as to make one's heart go pit-a-pat; suddenly, however, the fair one's thirst had to be quenched, and the readiness with which she handled the pewter and lowered the contents thereof, told us that she was not the fair damsel we thought her, but a he in disguise.[16]

Indeed, alcohol plays no small part in the tale of how the camp came to take place under canvas in the first place. Until 1880 the event had in fact been based not outdoors, but in Harrogate's highly reputable Commercial Hotel. That year, however, not only saw the arrival of many young cyclists, but also a troop of yeomanry (volunteer reserve cavalrymen), who were equally bent on enjoying themselves.

As might be expected, accounts of the riotous jamboree which occurred after the two groups came together that evening are somewhat sketchy. According to Lacy Hillier, events would not have spiralled out of control quite so quickly had it not been for 'a certain keg of whiskey', which a group of 'sundry Scotchmen from Edinburgh' made freely available in their hotel room. After full justice had been done to its contents, the evening culminated in a hotel-wide skirmish taking place between the cyclists and their yeomanry brothers. Henry Sturmey, another one present that night, recollected:

> Sundry boots were missing, and many others mixed. One valiant cavalryman's top boots were filled with water, and others were up the chimney, and were only discovered some considerable time afterwards. Even the sacred quarters of

the Colonel of the regiment were invaded, and that gentleman, we learnt, had valiantly drawn his sword in order to
repel the invaders… I distinctly remember one excited little
Frenchman, who declaimed loudly as to the disturbance of
his slumbers. 'I haf not sleep dis tree nights, and now – ver
are my poots?'

For those who participated in the mock battle, the whole affair acted as a
tremendous bonding experience. Sturmey described how C.B. Wilson
and A.B. Perkins, two cyclists present that evening, were discovered the
following morning

… calmly sleeping in the folds of a Union Jack, which they
had detached from the poles hanging out of their window,
and in a fierce struggle for the possession of the colours had
ended up rolling themselves up in it from either end, neither
party having been able to secure the victory.[17]

Not surprisingly, few others staying in the hotel were found in such
a peaceful and happy state the next day. With the raucous goings-on
making sleep impossible, the hotel's owner saw many of her regular
customers decide to leave and find lodgings elsewhere. This not only
ensured that those present were given an unceremonious send-off by
the landlady, but as stories spread to other hotel owners, they quickly
discovered that all Harrogate establishments would henceforth close
their doors to cyclists. 'Such were the circumstances which gave birth to
the Harrogate Camp,' concluded Sturmey, as members were left with no
alternative but to create their own accommodation and camp out.

Although they had been forced to move their event outdoors, it
was soon apparent that the freedoms created by this new arrangement

more than made up for the unseasonal rainfall for which the festival became renowned. From an initial attendance of around seventy clubmen the event continued to grow, and by the closing years of the 1890s was attracting cyclists from all over Britain and Ireland. Aside from the drinking and late nights, those who attended the event continued to delight in the warm and hospitable atmosphere they experienced amidst old friends and newly met kindred spirits. As we shall see, the abandonment of ordinaries in favour of pneumatic-tyred safeties during the 1890s did little to detract from these age-old attractions of club life.

These developments did, however, create a cycling scene which was far less dependent on the activities of organised associations. Improvements in machine design, not to mention rural areas gradually becoming more hospitable to visiting cyclists, meant the logic of safety in numbers during weekend excursions had now largely gone. Cycling's newfound fashionability also removed any stigma which had previously accompanied being one of the 'great unattached'. With its recognisable uniform, guidebooks and established network of inns and hotels, the Cyclists' Touring Club offered more appealing forms of status and security to those who wished to cycle with a select few friends and steer clear of the obligations which came with formalised club life.

The transformation of cycling into an easily accessible and everyday activity was thus not without its problems for clubs. With the pastime no longer firmly centred around their activities, during the nineties they had to cede much of the limelight they had occupied in their early years. Hampton Court and other large-scale meets were abandoned as popular attention shifted to Battersea Park and the sudden upsurge of female cyclists. And as their outward faces underwent a noticeable shift, so too did their internal dynamics and structures. For it soon became apparent that weekend runs on safety bicycles made for an altogether different group experience than those previously enjoyed on ordinaries.

'Again a British boy'

Amid all the furore and criticism which surrounded 'scorching' cyclists during the 1890s, there was one group who harboured a special grudge towards their fast-paced riding: members of clubs themselves. Alongside fearful road users and pedestrians, and middle-class riders keen to defend the 'respectability' of their pastime, clubmen often fiercely resented the accompanying presence of a 'scorcher' during their Saturday spins into the countryside. Articles in gazettes and the cycling press make it apparent that the high speeds attainable on safeties were placing a serious strain on the long-standing practice of clubs enjoying runs as one uniform body. Writing in 1893, a member of the Hackney-based Ariel Cycle Club complained:

> Long straggling lines of baked humanity, perspiring and miserable, running till they are exhausted, and affording fine scope for derogatory remarks to other users of the roads, take the place of the old-time Club Run, when we went out together, stayed together, and came home together.[18]

With the demise of ordinaries, whose height and cumbersome nature had acted as an effective leveller of pace when riding along poorly maintained rural roads, finding a speed which suited all those present on a run became increasingly problematic. While many complained about being forced to travel at rates they found uncomfortable, those keen to push and exert themselves could easily find riding alongside their more leisurely club mates frustrating and restrictive. 'The average club run is, in my opinion, a failure, and certainly is not conducive to health,' remarked one *Cycling* contributor, who characterised an average excursion as:

> A start at 2.30 p.m. – a 10 mile run – a drink – a lounge – tea – a sing-song – more drink – a start for home at 8.30. An outing lasting seven hours, of which two, perhaps less, are devoted to pedalling. Herein lies their idea of pleasure. Pleasure, forsooth![19]

These widely discussed difficulties were exacerbated by another consequence of the shift from ordinaries to safeties: clubs' changing demographics. Following the invention of the safety bicycle and pneumatic tyre, it was increasingly easy for 'old boys', who would have typically retired from high-wheeled riding in their thirties, to continue enjoying club activities into later life. Still attracting young men eager to make their names as racers, well-established institutions now found themselves home to a membership which, in terms of age at least, was highly diverse.

While senior members maintained a healthy level of physical fitness, these changing circumstances meant many clubs relaxed their older regulations. The Bristol B.T.C., for instance, split themselves into 'fast' and 'slow' divisions when undergoing weekend runs, taking different routes to a pre-arranged meeting point. Such an approach allowed the 'young man scarcely out of his teens' to indulge his urges to 'vie with his fellows, and to surpass them on the path, the road, or in other directions', while also satisfying older members for whom pleasure took an altogether different form:

To find oneself miles from anywhere, with blue skies over-
head, and the mossiest of banks inviting a loll, a smoke, and
a quiet chat with congenial clubmates – to experience these
things is to taste the real sweets of the cyclists' existence.[20]

Stanley Cycling Club members took a similarly laid-back approach
to club activities. 'Uncle', a frequent contributor to their gazette, stated his keen aversion to being a 'nasty fast young thing', and instead
enjoyed runs 'peacefully doddering in the direction of my beer'.[21] One
of his club mates described the start of the club's August bank holiday
tour in 1899:

The line of riders soon lengthened out, and I decided to stay
behind and look after 'Grandpa' while Scarfe and the 'Boy'
Ward 'put it through' Deaves all the way to Bury. The face
of Deaves, otherwise known as the 'Gay Lord Quex' when
he arrived at Bury, was the colour of beetroot warmed by the
midday sun.[22]

It is hard to see how well-established institutions could have survived
if younger members were denied the opportunity to 'put it through'
each other, or older members were unable to enjoy more peaceful
weekend dawdles. Nevertheless, such practices did not come without
their accompanying difficulties. After all, the friendships and close
bonds which had traditionally existed between clubmen were built
upon on the principle that they would share in the pleasures and
dangers of cycling. If members were going to spend much of their
time cycling in age-specific groups, maintaining these close-knit
relationships on a club-wide level suddenly became a much more
challenging proposition.

Weekend runs did, however, have the happy knack of bringing together those at contrasting life stages. For there was one aspect of journeying into the countryside which every member, no matter what their age or athletic ability, could agree on. Namely, the importance of stopping at least once over the course of a ride and fully indulging their appetite for food and drink. Pubs, inns, hotel restaurants and cafés acted as highly agreeable focal points for club excursions, providing plenty of opportunity for common enjoyment and relaxed sociability between members of different ages. Describing a recent ride he had enjoyed with his fellow club mates, one Argus Bicycle Club member reflected.

> There was much variety – the only points on which they resembled were the capacity for tea, bread and butter and eggs, and the universal animal spirits and good humour.[23]

Through providing a regular flow of hungry customers not seen since before the coming of the railways, by the 1890s visiting cyclists had done a huge amount to bring prosperity to country establishments. Articles in club gazettes make apparent how entertaining even a small number of riders enabled proprietors to rack up a healthy day's takings. Detailing a tour he had undertaken with three other club mates, which included a particularly ravenous individual named Charles, one Stanley Cycling Club member recounted how, during a stop at a hotel restaurant:

> They brought in a large joint of beef; Charles complained that all the goodness had been cooked out of it, but had two helpings. They brought in a large joint of cold veal. Charles complained that the person who cooked it ought to

be poleaxed, but had a plateful and a follow. They found a cold fowl, which Charles declared had been a professional gymnast. He ate a third. Then they brought us ham. Charles swore it was rancid, but cut and came again more than once.

With Charles still not completely satisfied, the group proceeded to finish off 'a large gooseberry tart, and some cheese, and a pint-and-a-half of beer each'. Providing a short summary of his touring partners, the author of the piece concluded:

> Will Futcher can peck a bit, 'Betsy' can hold his own, I am a wastrel, but, weight for age, and he says he is 56, I would back Charles Young at eating against the world for millions. We started for home at once... full of beer and beef.[24]

The pleasure the Stanley clubman took in witnessing his older club mate clean out the hotel pantry also brings to life the 'universal animal spirits and good humour' that united clubmen young and old. That such relationships existed can initially come as a surprise. The abandon with which Charles cleared out the hotel pantry certainly stands in stark contrast to conventional images of older middle-class men from this period. As fathers and senior workplace employees, in their day-to-day lives they would have been required to act as respectable authority figures, much like the one portrayed in the piece 'Men I Have Toured With', published in the Cyclists' Touring Club monthly gazette. Described as 'The Elderly Young Man', he was introduced as:

> In his fifties, has a very solid wife who wears the breeches (not to be confused with the rational costume!) a daughter who

will not see twenty-five again, and is in business a very stern and unrelenting solicitor. I have even heard him described as a narrow-minded prig by non-cycling acquaintances.

As hinted by the title given him by the author, however, once out on his machine he swiftly abandoned contemporary expectations of how men his age should behave. Observing that 'his memory would appear to have skipped the last twenty-five years of his life', the writer invited his readers to:

> See him at the village fair, at the shooting gallery, smashing pipes and cracking jokes with the circus clown, or in the New Forest climbing a tree after a carrion crow's nest, or in the Lake Country scampering down a fell side, or in the South Coast doing marvels in fancy diving from the pier head, or abroad playing the fool with a pompous *gend'arme*, and you would think him his daughter's younger brother, and you may be sure that she wouldn't own him.[25]

Reading contemporary accounts, it soon becomes apparent that 'Elderly Young Men' such as this were far from an isolated phenomenon. 'Cycling seems to possess a potent and peculiar charm to the middle-aged, aye, and even the elderly man,' observed a *Cycling* article titled 'It Makes You Young Again', which went on to comment:

> It is no uncommon thing to see the man of forty, in company with a party of younger men, acting in a manner that seems positively childish, when considered, though, perhaps, at the time the circumstances would hardly warrant your thinking so. We have often noticed this levelling influence of the sport,

and when you see bearded men – rulers among men, we may say – beyond the prime of life, vaulting five-barred gates, turning somersaults, and otherwise sacrificing the dignity and discretion that is generally supposed to pertain to age for the frolicsomeness of youth, you cannot help believing that cycling does in reality give man a new lease of life.[26]

Cycling was undoubtedly right to recognise the role the bicycle played in bringing out these changes. By taking older men away into new environments where they were not defined by their day-to-day identities, it made it far easier for them to abandon 'the dignity and discretion which is generally supposed to pertain to age'. That the exhilaration brought about by cycle riding also helped them cast off the years can be clearly seen in a piece by 'The Unchained' in the Bristol B.T.C.'s gazette. A self-pronounced 'boy of the old brigade', he celebrated how cycling in South Devon allowed you to

> … forget your cosmopolitanism and every other ism – again a British boy and proud of it, remember Nelson and Wellington and the brave tars and soldiers, who prevented the Corsican usurper from ever planting his foot on old England's shores. Presently you'll find yourself humming, strumming or shouting as of yore – Two skinny Frenchmen, One Portuguese, One jolly Englishman, Can lick 'em all three.

Further on in the piece and still in full 'British boy' mode, he described how when visiting sites from where 'old England's watch dogs' sailed out and 'singed the King of Spain's whiskers' when the Spanish Armada threatened, you could

> ... in imagination become a boy once more, and again
> experience that exultation that fired you when you first read
> 'Westward Ho', and when you took two slabs of wood and
> a tintack and made a sword with which you slew scores of
> moustachioed Spaniards, and rescued fair damsels and
> countless treasures (not forgetting the treasures) from their
> clutches.[27]

While he spared little detail in describing his childhood regression
to other members, quite how much 'The Unchained' revealed to
passers-by remains unclear. Although he could be sure no one would
recognise him far away from home, one assumes that his boyish
mindset was mostly lived out 'in imagination'. He would surely have
known that for anyone who happened to witness it, the sight of a
solitary middle-aged man singing and shouting as he cycled along the
cliffs of South Devon would, to say the least, have been something
of an oddity.

From this perspective, we can recognise just how important the
presence of fellow club mates was in aiding the transformations
described above. Within associations where age took a back seat in
defining clubmen's relationships and was replaced by the sense of
shared identity which came with cycling together, older members were
liberated from home and work identities. Common desires to make
merry and have fun ensured that those around them not only accepted, but eagerly supported the processes by which they shook off the
restrictions of age. Seemingly immune to the effects which might be
expected following their midday meal, the article detailing the Stanley
Cycling Club tour described how after lunch 'the old 'uns – I refer to
William and Charles – began to show rare sprinting powers'. As told
by the writer, this was something that

… only required a little encouragement. I rode alongside Charles and hinted that William fancied himself at riding. Charles remarked that 'he could beat his head off.' 'Betsy' sidled up to William, and said that on the next level bit Charles had announced his intention of taking his number down. We gradually brought the two together, and taking a glance at each other, they were off for a couple of miles – the two 'who ride only for pleasure' were plugging away for all they were worth. I do not know how it finished; 'Betsy' and I laughed so much we could not keep up with them, but both admitted to me afterward that they did not think the other would 'come that game again'.[28]

Immersing oneself in articles such as this, one increasingly acquires a sense of just how important club life was to those who participated in it. Further to the simple pleasures of cycling, the fellowship members experienced within their institutions enabled them not only to rebel against the passing of time, but also to find ways of coming to terms with it. Writing before the start of the 1898 Harrogate Camp, J.B. Radcliffe, who belonged to the pastime's 'old school', reflected:

Time has sped swiftly away since the first embodiment of the Meet, and with its departure have passed from our midst… many of the choice spirits that were wont 'to set the table on a roar'. A few of us are left to tell the tale of the brave days of old, with their cherished memories, consecrated by the flight of the inexorable scythe-bearer. The milestones on the highway of life are rapidly being left behind, and as we speed on towards the end of the pilgrimage, and as the shadows grow longer, we approach the period of reflection. We cogitate upon the past, its incidents, its lights and its shadows.

Not getting too disheartened, however, Radcliffe went on:

> It is not a dull, lifeless, apathetic reflection. No, there is much delight in dwelling upon old friendships, old comradeships, and on those happy days when care and its attendant heart-depressing sprites fled like shadows before the sunshine of youth and the sturdiness of early manhood. Of all the things in this world, cycling and camp life are hopeful, buoyant, and healthful.[29]

For all their heartfelt and delightfully written articles, however, it must also be noted that clubmen did also produce pieces that make for altogether more uncomfortable reading. Keen to remain male-only institutions, many articles display the conservative attitudes to female cyclists we have already encountered. At a time when the British Empire was underwritten by a belief in the innate superiority of the 'British race' to colonial peoples, smoking concert songs mocking 'niggers' also bring to light deeply entrenched racist views which were common in this period.[30]

Little is gained by making hard moral judgements on those who lived in a society far removed from our own. Nevertheless, the ways in which we look at history will nearly always be informed by the standards of today. Values expressed by supporters of rational dress, for instance, are ones which are now firmly established. As such, we are especially interested in how individuals such as Viscountess Harberton fought against what are now out-of-date attitudes, and sought to move society in the direction of the present.

The chief problem with such an approach, however, is that we can very quickly skew and distort the past. Looking back from the perspective of today can create the impression that societies were split along the lines of 'good' progressives who fought for modern rights and free-doms, and 'bad' conservatives who sought to deny them. As discussed in connection with Harberton and the Rational Dress League, this not only risks simplifying those who were in some but not all respects forward-thinking characters into straightforwardly 'modern' figures. It can also define many others solely by the fact that they stood in the way of 'progress', portraying them as bigots or relics beyond our compre-hension or understanding. And if we are being honest, the privileged middle-aged clubmen we have just explored will nearly always fall into this latter category.

From this perspective, the activities of 1890s club cyclists are an especially rich resource for making sense of a world very different to our own. The gazettes they produced make it impossible to ignore the basic humanity which connects us to all of those who lived through this, or indeed any other historical period. In forcing us to recognise this, they provide a view of the past which is richer, more colourful and ultimately far truer to life as it was in fact lived.

The arrival of the ladies

For all the conclusions we may draw from their writings, it would be a mistake to focus entirely on the select few clubs whose activities have been explored above. Wonderful resources though they may be, gazettes were by and large the products of wealthy, well-established institutions, whose activities extended back into the days of ordinaries. Made possible by access to the capital necessary to print and freely distribute monthly newsletters to their memberships, these official publications chimed with their acquired sense of self-importance. The Bristol B.T.C. was one organisation that listed the 'high position and

16 Photograph of four club cyclists relaxing with their safeties (date unspecified but believed to be early 1890s).

17 A large group of mixed-sex riders enjoying themselves during a mid-ride break (date unspecified).

influence of the club, its solid foundation, its old establishment and its alertness to pioneer any movement for the benefit of cycling and cyclists generally' as all being 'weighty arguments' for publishing the thoughts and opinions of its members.[31]

By the 1890s, however, older institutions such as this were being increasingly dwarfed by the many others that had sprung up following cycling's sudden surge in popularity. Alongside the Bicycle and Tricycle Club, by the close of the nineteenth century Bristol was home to thirty-four other cycling clubs, ranging from 'Crusaders' and 'Jockeys' to 'Bohemians', whose name was surely most in keeping with the city they belonged to.[32] Other major urban centres were similarly inundated. 'Never in the history of cycling was there such a long list of club runs by Leeds cyclists as that published today' proclaimed the city's *Evening Express* at the start of the club run season in 1898.[33] After extensive research, Henry Sturmey, whose writings extended well beyond lively reminiscences about early club life, listed 1,816 cycling clubs as being active across Britain in this same year, consisting of some 130,000 members.[34]

Within this vast array of new institutions were a few spawned by cycling's newfound fashionability. Possessing an exclusive membership of society bikists, the activities of the Trafalgar Bicycle Club in London took place on a private banked track in Trafalgar Square, West Kensington. Not only did this allow upper-class riders to avoid the crowds eager to watch their activities in London parks, but an adjoining club house also included dressing, drawing and dining rooms, as well as a spacious balcony where 'in summer, luncheon and other meals can be served'.[35]

Evidently this was not a typical organisation. Most clubs continued to serve their age-old function – allowing like-minded individuals to socialise and share in the enjoyment of weekend runs into the countryside.

The names of 1890s cycling clubs highlight how they often grew out of pre-existing social groups and networks, for whom cycling had now become a shared enthusiasm. Many had links to churches or Christian charities, with the town of Darlington home to at least four such institutions: Darlington St Cuthbert's, Congregational, Band of Hope and St Hilda's. Works-based organisations were also common, with institutions such as the Surrey Commercial Docks Cycle Club and Thames Ironworks Cycle Club highlighting how, with falling bicycle prices, members of the working classes were increasingly entering club life.[36]

A typical example of one of these new types of club is the Ancoats Wesleyan Cycle Club from Manchester. Established during the 'craze' in 1896, it was set up and run by members of the local Presbyterian Church with 'Reverend S. Marriott', whom one assumes practised as their vicar, serving as club president. The club's rule book and committee meeting minutes show that the club's weekend runs maintained older traditions, being enjoyed in an orderly and respectable manner. As well as wearing an official club badge and hat, members continued the practice of cycling behind a Captain, who had 'entire control of the club runs and meets'.[37]

In other respects, however, the club was markedly different to those that had preceded it. With no interest in training for race meets, members enjoyed their Saturday excursions at a leisurely pace, rarely travelling more than fifteen miles. Rather than camps and boisterous overnight tours, the club's stand-out social event was an organised picnic, enjoyed alongside fellow churchgoers and wider 'friends of the club'. And unlike most well-established institutions, as women's participation in cycling became ever more widespread, its all-male membership was soon subject to a noticeable transformation.

Up to the mid-nineties, cycling club life had been a male-dominated affair. While female tricyclists were invited into some institutions, clubs reflected cycling's status as a primarily masculine pastime. As indicated by the Stanley Cycling Club's rule that 'no dogs or ladies' were to be admitted onto club premises, this was something that most clubmen were more than happy with.[38] For although female friends and relations were occasionally invited to participate in a 'ladies' day' when they joined members in picnicking in the grounds of a country house or a similarly 'respectable' activity, women's involvement in club life was widely seen to pose an 'insidious danger' to group harmony.[39]

Some prophesied that when faced with the temptation of lady members, club mates would quickly abandon each other's company to spend time with their new recruits on a more intimate basis. One Argus Bicycle Club clubman feared that female accompaniment on their rides would mean waving 'good-bye to our merry teas and happy evenings, for many fellows would be seeking out the quiet secluded nooks beneath the trees, and I fear it would be very much a case that "two's company, three's none".'[40] Others were quick to recognise how travelling alongside 'ladies' would prohibit members' exuberant activities, forcing them to behave in a far more restrained and 'respectable' manner. A poem in the gazette of the Stanley Cycling Club observed that for their average club mate,

> When a ladies' day draws near,
> He leaves off smoking, gives up beer,
> And dons immaculate Sunday gear,
> All for the sake of the ladies.[41]

Horrified at the thought of living out such an arrangement on a weekly basis, the Stanley was one of many older institutions whose doors remained firmly closed to female cyclists throughout the 1890s. Although others allowed in female members, this was typically through a separate 'ladies section', whose runs took place separately from their male counterparts. While the two sections would often come together for evening and weekend social activities, this arrangement was agreeable to male members keen to 'scorch' and indulge other urges when out on weekend excursions.

On a broader scale, however, there was a general movement as the 1890s progressed for clubs' weekend runs to take place as a part of a mixed-sex group. For newly established organisations such as the Ancoats Wesleyan Cycle Club, whose rides took place in a well-mannered and leisurely manner, the presence of female cyclists did far less to change the nature of club activities. After the notion that 'ladies be admitted as members and be entitled to any and all benefits' was unanimously passed in 1897, several female cyclists quickly became a part of the club. This development was mirrored by numerous other institutions nationwide, as the *Leeds Evening Express* noted the 'tendency nowadays to admit lady members to the "active" memberships of cycling clubs' the following year.[42]

Such developments were notable not just within the context of cycling club life. With contemporary social conventions making it far more acceptable for men to spend extended periods of time away from the home, not to mention strenuously exert themselves, for most of the nineteenth century, sports clubs had existed as 'male preserves'. Alongside lawn tennis associations, cycle clubs were some of the first to reflect the gradual loosening of older traditions, as men and women began to spend their free time engaging in more modest forms of physical activity together.[43]

Cycling was unique, however, in the extent to which it encouraged sociability between the sexes. As an exhilarating activity which took place away from watching spectators, the way in which it softened and freed up relationships between men and women is beautifully brought to life in photos such as the one on page 127. By enabling such relaxed and happy scenes to be lived out on a weekly basis, clubs that opened their doors to both sides of the gender divide embody David Rubinstein's assertion that cycling was something which 'brought the sexes together on more equal terms more completely than any previous sport or pastime'.[44]

But when we move beyond these photographs and examine the broader structures of mixed-sex clubs, notions of gender equality quickly disappear. Although male members were happy to invite women into their institutions, overall running of clubs remained firmly in male hands. Committee members who planned weekend runs and other activities were nearly always exclusively male, as new female recruits found themselves barred from central decision-making.[45] While organisations such as the Hull St Andrews Cycle Club may have hinted to prospective members that 'who knows, you may next year elect a lady captain', a female cyclist leading and instructing her male companions was unthinkable. Condescending and authoritarian attitudes are very much in evidence in another piece in the club gazette, titled 'Cycling Hints for Ladies', published at a time when it had just opened its doors to women, which provided the following advice:

> WEARING A VEIL – A Veil is a good thing in hot and dusty weather, is a good protection against dust and flies, and tones the face down nicely.
> TO AVOID GREASING YOUR SKIRT – After oiling your machine down, rub off any oil from the outside bearings.
> STYLE – Cultivate a good style of riding, sit upright and don't stick your elbows out, nothing looks so objectionable as the latter.[46]

It comes as little surprise that the Hull St Andrews Cycle Club continued as a solely male institution. Even after warning potential recruits that 'independent of man as you may be, you would but cut a sorry figure a dozen miles from a railway station with a punctured tyre, and these things will happen you know', members were left puzzling as to why they had been unable to gain a single female club member. If they had looked widely enough, however, answers were at hand. For women were implementing the obvious solution to having a greater say in the running of their clubs – bypass the men altogether, and set up their own.

'Our girls a-wheel'

As with nearly every other aspect of female cycling, women-only clubs were given an enormous boost by the end of century craze. While these institutions had previously been rare, from 1895 onwards they underwent a dramatic upsurge in popularity. As well as becoming increasingly apparent in Scotland, Wales and Ireland, 'ladies' clubs' were especially common in England. By 1900 over a hundred organisations existed nationwide, active in virtually every major English city, town and region, and were particularly noticeable in and around London.[47]

Much like rational dress or racing, the activities of female-only associations ensured the wider public took an even greater interest in women's cycling. In keeping with older traditions, runs typically took place behind recognised Captains, with all members wearing club badges and attiring themselves in a common outfit. Combining the latest female fashion trends with historically male practices, they acted as highly conspicuous symbols of women's increasing entry into the pastime. According to a piece in *Hearth and Home* on the first meeting of the North Somerset Ladies' Cycling Club:

> All the ladies got away well together directly the captain's whistle was sounded, and the large and fashionable crowd assembled generally expressed their admiration at the riding. The captain was mounted on one of Singer's Grand Modele de Luxe cream and gold machines… The colours of the club are black-and-white ties and hat-bands, and the badge is a cycle wheel with monogram, 'L.C.C.' [48]

Cycling as an all-female group not only provided a means of showing off to 'large and fashionable' crowds, it also brought a wider set of benefits when members reached out-of-town settings. At a time when lone female riders often received unwanted comments from other road users, cycling with fellow club mates brought the promise of safety and mutual support. It is doubtful that many members of the South West Bicycle Club, whose runs to Brighton saw them arrive back in Tooting at midnight, would have taken night-time journeys without the protection of other riders. [49]

For those eager to challenge the wide range of restrictions faced by female cyclists, it also made perfect sense to band together and try to force changes as a collective. Several clubs, predominately based in and around London, enjoyed close links with The Lady Cyclists' Association (L.C.A.) and Rational Dress League, with it being common for leading figures from these institutions also to possess local club membership. Members of the Western Rational Dress Club attended the League's widely publicised ride from London to Oxford in 1897, as they sought to bring about the following objectives:

1. To promote a dress-reform whereby Ladies may enjoy out-door exercise with greater comfort and less fatigue.
2. To advocate the wearing (particularly for Cycling) of the Zouave or Knickerbocker Costume, as adopted by the ladies of France, Germany and America.
3. To take all necessary steps, in connection with kindred Associations in London, to encourage this desirable reform.[50]

Alongside efforts to promote rational dress, there were also female clubs that encouraged women's racing and competition. The Chelsea Rationalists were especially notable in this regard, organising their own race meets with club captain Monica Harwood winning the 1895 six-day event at the Royal Aquarium.[51] Although not all who support-ed dress reform took such a positive view of women's cycle racing, the cluster of clubs associated with the L.C.A. and Rational Dress League were united in their efforts to break down barriers faced by female cyclists. Writing a piece titled 'Girls' Cycling Clubs and Associations', clubwoman and L.C.A. member N.G. Bacon proclaimed:

'Unity is strength!' With such a motto for our foundation stone, it is wise to discuss the advantages to be gained by Our Girls A-wheel with co-operation and organisation… In the cycling realm it is but natural that our girls should discover that they are capable of enjoying both individual and collective life. Club life… teaches first social opportunity, then friendship and the advantages of unity.[52]

A passionate advocate for female cycling, Bacon was one of many who celebrated the growing numbers of clubs run by and for women.

While not all promoted political agendas, in whatever form they took these new institutions carried the promise that 'a new era has commenced, and the womanhood of the twentieth century may be the eyewitness of a great revolution'. Unlike mixed-sex organisations, they were not just symbols of female participation in a pastime that men continued to preside over. Instead, they claimed for women the long-held masculine privilege of establishing their own independent associations. As such, these clubs were sustained by the same belief that had made rational dress and racing so controversial – that women could freely participate in the pastime in the same manner as the opposite sex.

It comes as little surprise, then, that early female organisations were subject to widespread worries and criticisms. Faced with their development, the cycling and wider press were soon seeking a means to play down the principles that underlaid them. Their most widely used tactic was the production of fictional sketches which mocked the idea that the 'fairer sex' were capable of, or even really wanted to, organise and carry out their own weekend excursions.

In the cartoon opposite, from *The Graphic*, the rationally clad members set off with earnest intentions to obey their rule that 'no gentleman was to be spoken to during our runs, or under any pretext whatsoever'. This decree is soon violated, however, as they are forced to seek male help when finding themselves unable to fix their Captain's broken machine. Upon meeting these same cyclists in a hotel restaurant (where else?), order is quickly restored as they fall for the charms of their male counterparts, who in a later development are invited into the club and bring a swift end to its independent existence.

In similar vein, a *Cycling* article titled 'That Horrid Cow!' detailed the misfortunes which befell the 'Advanced Women' of the fictional 'Seaton Ladies Cycling Club'. Home to members who 'worked hard

18 Print in *The Graphic* portraying a run undertaken by members of a
Ladies' Cycling Club (1895).

for Woman's Suffrage', held 'peculiar views on marriage' and 'publicly smoked cigarettes', their ride begins with them happily discussing leading female novelists, social reformers and journalists, with it being agreed that:

> All this nonsense about feminine inferiority is mere masculine invention. Our education, our habits, our code of morals, our very dress, have hitherto been modelled on arbitrary principles invented by man for his own purposes.

Their earnest discussions are soon ended, however, by an encounter with a cow in the middle of the road. Faced with this unexpected development, the writer had the lady members drop their machines and disappear behind a nearby hedge, from where they 'tremulously peeped through at the enemy'. Fortunately, a male cyclist again appears to provide help, shooing away the animal and rescuing the female club members from their predicament. The piece finishes with them remounting, and as soon as 'their deliverer had disappeared round the next bend, the Seaton Ladies C.C. cheerfully resumed the subject of the Supremacy of Woman!'[53]

The contrast between female-only cycling clubs and women's supposed dependence on male advice and support was irresistible to editors eager to mock the idea of these institutions. As the years progressed, however, these articles became less and less common, before disappearing altogether. For with the rapid growth of female cycling clubs, they became ever more impossible to reconcile with reality.

Admittedly, this was partly because there were only a select number of female clubs who, outwardly at least, strongly pushed against the status quo. While some promoted more radical agendas, there was a widespread commitment to affirming the 'respectability' of women's cycling. As with male institutions, ladies' clubs invited influential and well-regarded local figures to serve as club presidents. The *North-Eastern Daily Gazette* detailed how 'Mrs Stewart', wife of the Mayor of Darlington and a similarly 'staunch Conservative', acted as patron for the town's female cycling club.[54] With such an individual at its head, the Darlington Ladies' Cycling Club could hardly be portrayed as an institution which threatened the established social order.

At the same time, the success enjoyed by female clubs clearly discredited the notion that women could not run their own institutions. Sources such as the 1898 rule book of the Manchester Ladies' Oxford Cycling Club demonstrate how they successfully utilised the same structures as men's clubs, with a captain, sub-captain, treasurer, secretary and committee overseeing the club's day-to-day activities. Scheduling two club runs a week, on Wednesday and Saturday afternoons, members typically travelled just under twenty miles, a similar distance to many male organisations. There were also two longer-distance rides of forty miles to Liverpool and Chester over the course of the year, as well as an Easter weekend trip to Blackpool, of which further detail is sadly unforthcoming.[55]

In a society whose older generations had grown up with the belief that a woman's place was firmly in the home, associations such as this were a powerful symbol of changing social practices. With all officers prefixed with the title of 'Miss', one assumes that most members were young, aged somewhere between the teens and late twenties. At a time when middle-class women of a similar age were enjoying increased access to education and making their presence felt in the workplace, they symbolised growing understanding of female potential and capabilities. As Bacon put it later in her article:

> Let us for a moment reflect what an amount of charac-
> ter-building of the finest description is silently going on during
> these club rides… The sensitive steed exacts the attention of
> both the mind and the body… The brain, no less than the
> eyes and muscles of the body, comes into sympathetic action
> with the movements of the sensitive vehicle beneath the rider,
> who thus is unconsciously drawn away from herself… whilst
> she is at the same time being inspired by the most glorious
> sense of comradeship, which is the slender tie worthy of
> joining all sorts and conditions of our girls together.[56]

The 'silent' processes described by Bacon, and the way they developed members' minds and bodies, may well be where female clubs
achieved their greatest impact. As with rational dress and women's
racing, they were relatively short-lived phenomena, with few surviv-
ing beyond the turn of the century. It is hard to imagine, however,
that the autonomy and sense of self-reliance they provided to the
many who joined them disappeared with their decline in popularity.
Looking forward a few years, it is telling that groups of female cyclists
were heavily involved in the campaign tactics of the suffragettes, as
brigades of 'Cycling Scouts' travelled out into the countryside to
'deliver the gospel of votes for women'.[57]

Indeed, for the founders of the Women's Suffrage and Political
Union, the idea of using the bicycle to spread their message was an
obvious one. For during the 1890s the Pankhursts were members of
another institution that, through a national network of clubs, had
utilised much the same practice. No account of cycling club life in this
period would be complete without the story of how these organisations
were at the heart of a movement which sought to deliver the gospel of
another radical new concept: socialism.

Propagating good fellowship

It says much for the bicycle's utility that while the aristocracy were promenading around London's parks at the height of the craze, it was becoming associated with a very different type of inner city movement. With the renewal of ideas which had faded since the Chartist movement in the 1850s, the closing years of the nineteenth century saw a major revival in British socialism, attracting committed advocates in figures such as the artist William Morris, who was involved in both the Social Democratic Federation and the breakaway Socialist League. The Independent Labour Party was founded in 1893, as trade unions looked to secure the votes of recently enfranchised working-class men and gain representation in Westminster.

None of these associations, however, enjoyed the same popular influence as the one which most fully embraced the symbolic and practical significance of the bicycle. Unlike anything which had come before or indeed since, *The Clarion* newspaper inspired a populist movement which reached a huge predominately working- and lower-middle-class audience in this period. At its head was Robert Blatchford, a former solider and army sergeant turned journalist, who was converted to socialism after witnessing the appalling conditions faced by Manchester's poorer inhabitants. After producing impassioned articles about slum life for the city's *Sunday Chronicle*, in 1891 he left the paper to set up *The Clarion*, a penny paper through which he could freely express his politics. Blatchford used his first leading piece to proclaim:

> *The Clarion* is a paper meant by its owners and writers to tell the truth as they see it, frankly and without fear. *The Clarion* may not always be right, but it will always be sincere. Its staff

do not claim to be witty or wise, but they do claim to be honest. They write not for factions; but for the people... The policy of *The Clarion* is a policy of humanity, a policy not of party, sect or creed; but of justice, reason and mercy.[58]

With its first edition selling 40,000 copies, *The Clarion* sustained its early popularity by successfully carrying out these ideals. Alongside political pieces written with a keen social awareness, plenty of space was left for articles which emphasised the importance of 'loving one another as brothers and sisters' and sought simply to entertain the paper's readership. As Denis Pye puts it in his official history of the Clarion Cycling Club:

> One advertisement for the paper declared: 'There is nothing like it. There never was anything like it. There never will be anything like it.' And the reason why this was no empty slogan is that the *Clarion*, unlike other Socialist papers, espoused a Socialism which was not in the least solemn, difficult, highbrow, dreary, theoretical or dogmatic, but rather a way of life to be enjoyed here and now, in which men and women, young and old, would live in fellowship with each other in their everyday work and leisure activities.[59]

Sharing these principles, readers of the paper wasted little time in coming together and forming their own associations through which they could live out and spread *The Clarion*'s ideals. In the 1890s, promoted through the paper, various Clarion sporting institutions were founded, such as swimming and rambling clubs, as well as artistically minded dramatic societies and vocal unions. In keeping with the times, however, by far and away the most popular organisations were cycling clubs.

Tom Groom, who alongside six other Birmingham-based enthusiasts founded the very first Clarion cycling association in 1894, would later look back and reflect:

> By the end of the first year, over eight Clarion Cycling Clubs had been formed, and the next season started with over one hundred and twenty clubs. Within three years the membership had gone up from seven to over seven thousand. And the original Object of the Club: To propagate Socialism and Good Fellowship, was actively and vigorously pursued.[60]

The ways in which Clarion cyclists went about pursuing these twin objectives varied considerably. Most famously, members used their machines to broadcast the Clarion message to more remote rural audiences. When out on runs they gave open-air talks and addressed public meetings, while also distributing copies of *The Clarion*, leaflets and penny editions of 'Merrie England', a widely read pamphlet in which Blatchford laid forth the basic principles behind his socialism. In the build-up to the 1895 general election, country walls, fences and even trees were pasted with socialist stickers and slogans, placed there by enthusiastic Clarion cyclists. With socialist candidates eventually receiving just 45,000 of the two million votes cast, however, Blatchford was one of many unconvinced by this scattergun approach.[61]

But while it was said that 'a Clarion cyclist was a man with a bicycle – and a saddle bag full of "Clarions"', cycling's value to the movement extended well beyond the spreading of propaganda.[62] For Blatchford, who was inspired by Morris's assertion that socialism was the politics of beauty, the pastime spoke to his vision of society free from the squalor and pollution of industrial heartlands. As he put it in an 1895 article:

I never wander in the stuffy sordid London streets nor in the squalid gruesome Northern slums, but I think of the dancing sea waves, of the flower starred meadows and the silky skies of England. And by the same token the sweet air and sunny landscapes and still green woods bring up before my eyes with painful vividness the breathless courts and gloomy lanes, the fever beds and vice traps of horrible Liverpool, and horrible Manchester, and horrible Glasgow.[63]

To experience the fresh air and beauty of the countryside was, in Blatchford's opinion, to acquire a sense of what a socialist society would feel like. By recruiting working-class members and allowing them to see the contrast between their home communities and open rural spaces, Clarion cycling clubs hoped to 'make socialists' of those who had previously been outside their movement. While such an expectation can appear fanciful, encountering the natural world undoubtedly had a profound effect on Clarion cyclists. One club member recounted how he had become a socialist in early life after studying butterflies, for 'the loveliness of the insects had bred him in a desire for beauty also in the human world'.[64]

Even more closely aligned to the principles behind *The Clarion*, however, was the 'good fellowship' inherent in cycling club life. At the movement's heart was a loyal and lively sense of belonging which, as we have already seen, was a defining feature of these institutions. Writers for the paper utilised the types of humour and in-jokes normally found in associations such as cycling clubs, using these as a means of building a functioning nationwide community through no more than a weekly paper. While Blatchford was widely known as 'Nunquam Dormio' (Latin for 'I never sleep'), other senior figures took on rather less intellectual nicknames, such as The Bounder (Edward Fay), Dangle (A.M.

Thompson), Mont Blong (Montague Blatchford) and Whiffly Puncto (William Palmer), while all who belonged to the movement were known as 'Clarionettes'. Writing in 1910, Albert Lyons offered this assessment:

> The Clarion Fellowship is absurd and inexplicable, and – wholly charming. It is Socialism – real Socialism – this bond of genuine sympathy and kindness which exists between the readers of the paper which Mr. Blatchford edits. A stranger entering almost any town in England has but to proclaim himself a *Clarion* reader… to be assured of welcome and hospitality in the houses of friends whom he has never seen before. This is not practical; but it *is* Socialism.[65]

In much the same way, cycling club life was understood as bringing socialism into being on a small scale. Creating 'bonds of genuine sympathy and kindness' with fellow riders both embodied the ideals of *The Clarion* and offered a means of building a socialist society from the ground up. In the words of Tom Groom:

> To get healthy exercise is not necessarily to be selfish. To attend to the social side of our work is not necessarily to neglect the more serious part. To spread good fellowship is… the most important work of Clarion Cycling Clubs. Then, perhaps, the 'One Socialist Party' would be more possible and we should get less of those squabbles among Socialists which make me doubt whether they understand even the first part of their name.[66]

With such aims in mind, much of Clarion club life closely resembled that of other organisations from this period. Like the Harrogate

Camp, 'Clarionette' cyclists from across the country came together once a year for a well-attended Easter Meet. Although it was not quite as boisterous, drinking and smoking concerts still featured strongly, as members enjoyed 'right merry evening[s] of good fellowship'. After a plain-clothes policeman attended the 1898 meet in Chester so as to 'nip sedition in the bud', he later reassured his superiors that 'if these chaps kill anybody, it will be from laughing'. True to this appraisal, 'Dangle' would later look back and reflect, 'Ah, those Clarion Meets! I look back and deliberately declare that I have had more heart joy of them than of any other event in my long life.'[67]

Although suggestive of all-male revelry, unlike the established institutions explored previously Clarion cycle clubs extended a welcoming arm to female members. You could hardly claim to be creating a more equal and inclusive society, after all, if you barred one half of humanity. Many were founded as mixed-sex institutions, and clubs that were initially male were quick to gain female recruits. Sylvia Pankhurst later recalled how they 'promoted a frank, friendly comradeship amongst men and women, then very much less common than it is today':

> Week in, week out, the Clarion clubs took hundreds of people of all ages away from the grime and ugliness of the manufacturing districts to the green loveliness of the countryside, giving them fresh air, exercise and good fellowship at a minimum of cost... At our journey's end was always an enormous shilling tea, in which phenomenal quantities of bread and butter and tinned fruit rapidly disappeared, then a walk around, and frequently afterwards a brief 'sing-song', sometimes joined by members of other clubs who had ridden that way.[68]

The idea that the humour and sense of belonging that people experienced in associations such as cycling clubs could be transposed to wider society was what gave the Clarion movement its tremendous energy and popularity. Unlike complicated economic theory, it was an easily grasped concept that anyone could experience and live out for themselves. As we have already seen, it was the levelling influence of spending time with fellow enthusiasts that enabled older cyclists to find such joyous liberation through their club's activities. There are obvious parallels between the piece below, written by an older Stanley Cycling Club member who attended the 1900 Harrogate Camp, and principles expressed in *The Clarion*:

> Social distinction, rank of fortune, vanishes as you take your place at the shrine of Bohemianism; equal in one accord – the desire to knock as much enjoyment as you can into four days of grace. Can it be wondered then that men grown grey, be-bearded solemn pards, faced with the inevitable scourge of time, take to Harrogate as a duck to water, and for a brief spell throw aside the decorum and dignity incumbent with their station in life, to revel as they did in the heyday of their youth, when the blood ran freer and the pulse beat quicker? Such enthusiasts can fully endorse the poet's couplet,
>
> > 'Tho age is on his temple hung,
> > His heart, his heart is very young.'[69]

Of course, it is highly unlikely that the writer and others who enjoyed Harrogate shared Blatchford's political philosophies. With one clubman arriving in a £600 motorcar (which on arrival, after over-zealously applying the brake, he flipped into a ditch and immediately wrote off), camp attendees ranked as some of the most prosperous individuals in the country. While they celebrated leaving 'social distinction' and 'rank of fortune' at the camp gates, they did so safe in the knowledge that they were temporarily, rather than permanently, leaving them behind. However strong a flavour of youthful rebellion might be carried in articles such as the one above, there is an underlying acceptance that once their 'four days of grace' was over, normal life would resume its course.

Nevertheless, the camp and club life more generally were hardly just sites of escapism. As is made apparent by the joyous accounts produced by older members, they carried an enormous emotional value to those who partook in them. And this could only have occurred by them enabling people to live out the ideals at the heart of *The Clarion*. Breaking down differences in age, class and gender, clubs enabled thousands to socialise and bond with those with whom, in everyday life, such relationships would have been impossible.

Even if limited to a few blissful hours on a Saturday afternoon, this chance to live a good life in the company of others, to connect and to relax, remains a form of socialism that society can hardly do without. Despite the distance in time that separates us, and the sometimes arcane and indeed preposterous rituals that the clubs embraced, this call to find meaning and happiness through community speaks powerfully across the years, connects past and present, and reminds us what really matters in life.

4

Romance, Romeos and Chaperones

On board a boat bound for America in 1891, Harry Dacre had the time to look back and take stock of his career as a music hall song-writer. After a decade of working in the trade, it is fair to say his output had been defined more by its quantity than its quality. In his first two years alone, he claimed to have written more than seven hundred songs, producing compositions at the rate of a ditty a day. Possessing, among other things, a bicycle and yet more original material, he was one of many travelling to the United States hoping for a new start.

It is safe to say things did not begin as planned. Even before he had properly set foot in America, Dacre had his mood darkened by a customs officer who gave him the unwelcome news that he would have to pay an import tax on his bicycle. Still, as his friend and fellow songwriter William Jerome jokingly pointed out, things could have been worse: if he had brought a two-person machine the rates would have been doubled. If Jerome's message was little consolation, the way in which he delivered it certainly was. For rather than using the word tandem, he chose instead the rather more lyrical 'bicycle built for two'. Writing with his usual rapidity, Dacre worked this line into a song which would, after thousands of earlier attempts, secure him his little place in posterity:

Daisy, Daisy, give me your answer do!
I'm half crazy, all for the love of you!
It won't be a stylish marriage,
I can't afford a carriage
But you'll look sweet upon the seat of a bicycle built for two.

Even with such promising material, Dacre initially struggled to find a singer to take the song onto the stage. Eventually, well-known music hall act Katie Lawrence agreed to take 'Daisy Bell' to England, where her success was by no means instantaneous. Lawrence would later recall:

> When I reached London, I sang it for four or five weeks, but it never seemed to catch on or go at all well. I sang it in the provinces with a not much better result and I made up my mind to drop it. I returned to London on the afternoon of Whit Sunday... Suddenly, whilst at the station getting together my luggage and putting it on a cab, I heard someone humming 'Daisy'. A few minutes later I heard it again and found it was all over London.[1]

After taking off in London, 'Daisy Bell' quickly snowballed into a huge hit on both sides of the Atlantic, enabling Dacre to set up his own publishing house and secure a comfortable existence for the rest of his life. Indeed, such was the song's success that its influence was soon being felt beyond the stage and music hall. Ever eager to find new ways of entertaining the public, cycle race meet organisers arranged special 'Daisy' races between tandem teams of male and female cyclists, while a lively crowd sang along to the now familiar tune played by a brass band.[2]

Away from the race track, Dacre's ditty was also of inspiration to young couples seeking to tie the knot. During this period there was a brief craze for 'bicycle weddings', in which the bride and groom rode to and from the ceremony on their 'bicycle built for two'. One newspaper report on such an event, titled 'Daisy Bell's Wedding', reported:

> A 'bicycle marriage' has taken place in the quiet Surrey village of Ashtead. Every member of the party went to church on a cycle of one kind or other… The bride and groom occupied a 'tandem,' the former, of course, in the front seat, and piloting her husband safely over the rough bits of road.

The piece acquired an increasingly surreal feel as it went on to detail the clothes worn by the bride:

> A fawn-coloured cycling costume, knickerbockers included, and her coiffure, from which streamed a white veil, was garlanded with orange blossoms. The bridegroom was attired in light coloured raiment somewhat similar to the bride's. In fact the orange blossoms were the only distinguishing mark between the two. The bridesmaids and the best man followed as fast as they could, the former wearing 'scorcher' breeches, presented by the bridegroom.[3]

While the events described may have been sourced from hearsay rather than reputable journalism, there can be little disputing the association which existed between cycling and romance during the 1890s. Not only did it inspire one of the most popular songs of the decade, it was referenced in all manner of other areas of popular culture. H.G. Wells and Emile Zola both used bicycle rides as settings for burgeoning romance, while numerous short stories in periodicals featured unsus-pecting couples finding love when out riding. A male and female cyclist riding along together was also a stock image in adverts ranging from cycle clothing to muscle rub lotions, whose posters offered the reassur-ing caption 'no stiffness here'.

To continue a theme from the previous chapter, the bicycle created a range of new possibilities to experience intimacy and build relation-ships. From labourers in isolated rural settings, to Princess Maud who, fresh from her travels around Battersea Park, grew to know her future husband Prince Charles of Denmark while cycling together in his home country, the bicycle contributed to the love lives of people across the social spectrum.

What exactly was it about cycling that made it such a great facilita-tor of romance? What new opportunities did it create? To understand this, the best starting point is not tandems, but machines from even earlier in cycling's history. For well before bicycles built for two, there had been tricycles which served the exact same purpose.

'The pleasures of cycling, quotha!'

Out of the wide array of cycles that appeared during the late nine-teenth century, there are surely none more striking or fantastic to look at than the sociable tricycles produced from the late 1870s onwards. Accommodating two riders who powered either a single or a double chain set, these machines' primary attraction lay in the fact that, unlike previous bicycle or tricycle models, people could enjoy their weekend excursions while pedalling alongside a fellow enthusiast. They were popular not only with affable club mates and racing cyclists eager to team up on record-breaking rides: photographs make apparent that a

19 Photograph of a couple a few days before their wedding day alongside friends, family and bicycles (1899).

20 Lady and gentleman pictured before setting off on a ride on their sociable (c. 1891).

significant amount of inter-sex mingling occurred atop sociables. As an article in *The Tricyclist* put it:

> No man is more popular in his set than he who owns a sociable tricycle. His lady friends are always at home to him, and whether for a short evening ride, or for a more extended trip, he never need despair of finding a pleasant companion.[4]

Popular with single men looking to become better acquainted with lady friends, sociables were also a boon for married men, helping them to avoid the significant risk of marital duties putting an end to their precious weekend excursions. The design of sociables enabled one rider, who was seemingly always the husband, to take primary responsibility for pedalling, allowing him to induct his partner into the joys that could be found through cycling. Such a character was the journalist and racing tricyclist A.J. Wilson, who in his book *The Pleasures, Objectives and Advantages of Cycling* cried:

> The Pleasures of Cycling, quotha! My good Sir – or my dear Madam, if I am fortunate enough to have a lady reader – you can have no adequate conception of the genuine pleasure experienced during a cycling trip until you come to participate in it yourself. Prithee, then, procure a tricycle upon which to commence gentle exercise; attempt not to do more than a mile on the first day, five miles the second, and the same the third.

For those still harbouring doubts about the pastime, Wilson offered the reassurance that 'this is just to accustom your limbs to the motion', as he then invited them to 'occupy the front seat of my tandem' ('tandem' being used to describe sociables during the 1880s). Using the original and, it must be said, rather entertaining literary device of taking his reader on an imaginative journey out of London with him, he offered this encouragement:

> Excepting up hills, I can very easily keep the tandem running at a good steady pace; and you may rest your feet entirely on the foot-rest provided for the purpose… I have adjusted your saddle to suit your height, and your handles are in a comfortable position; be sure you do not attempt to work hard, on this your initial journey, and I can promise you a treat. So—! Now we are bowling along the suburban roads …[5]

Although those who rode sociables may not have contributed equally to the pedalling, travelling together afforded husbands and wives a welcome opportunity to escape the constraints associated with domestic life. Sociable riding provided both precious intimacy and the chance to delight in the scenery they were passing through. The bond this created between couples is clearly evidenced in another piece by Wilson, written in the gazette of the Stanley Cycling Club when he was thirty-five:

> I enjoy an occasional club run and appreciate the gentlemanly sociability I meet with among fellow Stanleyites; I revel even in an occasional race, or race attempt on road or path; but these things are mere incidents – pleasant excitements of the moment; and year in and year out I find no form of cycling so quietly enjoyable as to trundle out the Tandem and take my wife for an impromptu, objectless, unconcerned dawdle through the bye-lanes.[6]

Not all who went tricycling together, however, did so with the same relaxed motives as Wilson and his wife. For some 'leisurely dawdles' were not the point: they sought speed and adventure. Leading the way when it came to adventurous tricycling couples were the American pair, Joseph and Elizabeth Pennell. After meeting in Philadelphia in 1881 they quickly formed a productive working partnership on the back of their respective vocations: Joseph as an illustrator, and Elizabeth as a writer. Growing ever closer, in 1884 they married and moved over to London, enabling Joseph to take up a job with *The Century Magazine*.

Possessing a common love of cycling, the pair also quickly moved on to producing books which, utilising their respective skills, vividly brought to life the events they had experienced when riding their sociable. This included both longer and shorter expeditions – *A Canterbury Pilgrimage*, written in 1885, was modelled on Chaucer's famous tale, and detailed a three-day excursion they had enjoyed from London to Canterbury. Later books, such as *An Italian Pilgrimage* and *Our Sentimental Journey through France and Italy*, published in 1887 and 1888 respectively, detailed the far more challenging long-distance tours they had undertaken across these two countries.

This collection of works, which quickly acquired a widespread readership, made the Pennells minor celebrities within the world of cycle touring. Their books did a huge amount to popularise this branch of the pastime both as an individual activity and as something which could be enjoyed by couples seeking adventure and excitement on their holidays. After chancing on a couple of unsuspecting fellow cyclists in a Yorkshire pub in 1886, Joseph listened with interest as they told him about 'those Pennells', most notably the curiously minded husband who 'went around with his wife or sister or something all over creation'. Elizabeth's diary also records how admiring fans frequently came to visit their house in London, such as 'Mr. and Mrs. Harold Lewis from Philadelphia' who were 'just back from a tricycle ride of 2,200 miles, beating our record all to pieces'.[7]

The recognition the Pennells enjoyed certainly did not stem from over-romanticised accounts of their tours. As would be expected, lengthy journeys on a large and cumbersome machine, which required them to do battle with the elements, dodgy roads and inhospitable landlords, were not always the most enjoyable of undertakings. After a day of cycling into a headwind in France, Elizabeth recalled:

> I was so tired! Every turn of the pedals I felt must be the last. And the thought that we should reach Cosne but to begin the same battle on the morrow, did not help to keep up my spirits. In vain I tried to be sentimental. For the hundredth time I admitted to myself that sentiment might do for a post-chaise [a horse-drawn carriage], but was impossible on a tricycle. And all the time J kept telling me that if I did not do my share of the work I should kill him.[8]

Although the Pennells occasionally bickered over their respective workloads, they avoided another cause of tension between holidaying tricyclists: the driving of the machine. Many sociables were designed so the rider at the front, most often a wife, was responsible for the manoeuvring of its considerable bulk. The fallings out this could cause were detailed in *A Canterbury Pilgrimage*, where after stopping at an inn they met another rider who complained that he

> ... had to sit behind his wife – she had to steer, and he would not be surprised if he were seriously injured, or even killed, before he got back to London... His wife, who had joined us

a few minutes before, here grew angry, and a slight skirmish of words followed between them: she reminded him of the dangers they had escaped through her nerve and skill; he recalled the dangers into which they had run owing to her thoughtlessness and timidity.[9]

The encounter with this character – pithily described as 'a short man with a bald head, who wore the Cyclists' Touring Club uniform' – was one of many delightful short sketches in the Pennells' books. One especially humorous scene was played out after a day of travelling through driving rain in France, which by evening had thoroughly drenched both the Pennells and their luggage. After they had made it to the town of Beaumont and stopped for the night at the first inn they found, a kindly landlady insisted the pair spend the evening by a fire in her private sitting room and change out of their wet clothes into some outfits she had dug out for them. These were not, however, the most becoming of costumes. As Elizabeth wrote:

> I flattered myself that I, in her neat wrapper with a little white ruffle in the neck, made quite a presentable appearance. J—'s costume, consisting of her husband's dressing-gown and a short kilt improvised out of a plaid-shawl, was more picturesque, but less successful.[10]

Although the landlady produced an illustrated history of London by the writer George Thornbury for their entertainment, 'we were more taken up in looking at each other, and were reasonably serious only when she was in the room'. Though hardly full of starry-eyed sentiment, Elizabeth's quietly affectionate descriptions of their tours, which often feature Joseph swearing, struggling to communicate with locals, and not infrequently combining the two, make apparent the close ties that tricycling had built between them.

The arrival of the safety bicycle and pneumatic tyre certainly did not put an end to husbands and wives going on journeys together. Indeed, riding along separately on two pneumatic-tyred safeties would soon prove itself to be a much more comfortable, practical and, by the mid-years of the 1890s, cheaper activity compared to travelling atop a sociable tricycle. This is reflected in the activities of the Pennells who, by the time of writing *Over the Alps on a Bicycle* in 1898, had abandoned their old machine in favour of two up-to-date models.

Even more significant, though, was their impact on those yet to be married. The rapid uptake of the new safeties by both men and women created a wealth of new opportunities for sparking the fire of romance. Well before the internet, smart phones and online dating, the bicycle was a new technology that brought people together in previously unimaginable ways. And the consequences of this were felt in all manner of unexpected places.

Men, women and gene pools

As with all other aspects of the pastime, it is an unfortunate fact that our picture of the bicycle's impact on the relationships between late nineteenth-century men and women is heavily skewed towards more affluent individuals who came from urban environments. Those who produced written sources, such as Wilson and the Pennells, were invariably well-to-do townsfolk who enjoyed cycling as a leisure activity. In contrast to the abundant documentation produced by members of this social group, there is a sad dearth of first-hand accounts on how cycling benefited Britain's rural communities.

This is particularly regrettable given the unique potential of cycling to transform the lives of poorer countrymen and women. At a time when nearly all the journeys they undertook were made on foot, owning a personalised means of transportation for the first time must have been world-changing. A bicycle would have brought with it the opportunity to find employment which was previously inaccessible, reach what had been remote towns and villages, and, of course, build relationships with the new and unfamiliar people you met there.

Although written sources are scarce, we can occasionally glimpse how the extra mobility offered by the bicycle helped transform late nineteenth-century rural life. Particularly useful in this regard are Church of England marriage registers, which record the names and residences of couples who wed in this period. With marriage of course dependent on the bride and groom previously spending time together, their home addresses provide telling evidence for how far people usually travelled beyond their local communities. By examining how these distances changed over time, we can acquire a picture of how the bicycle may have brought together those from previously isolated villages and hamlets.

Fortunately for the writer of this book, others have already invested the time and effort necessary to undertake such an analysis. In an academic article published in 1969, the geographer P.J. Perry shared his analysis of the registers of twenty-seven Dorset parishes in the hundred years following 1837, and identified trends which provide compelling evidence of the impact of the bicycle. For from 1887 onwards, a period which closely aligns with the invention of the safety bicycle, there was a significant increase in marriages between individuals from previously disconnected parishes. As Perry himself put it:

Before the coming of the bicycle the countryman generally travelled on foot; carrier's vans and carts, the railways, and other possibilities were too expensive for general use and often not in accord with the needs of the would-be traveller. Dependence on walking, however, much restricted the area of frequent and everyday contact... It was this situation that the bicycle, *inter alia*, transformed, although exactly how and when remains uncertain.[11]

Though we are lacking specific details, much can be said in support of Perry's conclusion. The price of a second-hand safety or out-of-date ordinary certainly would not have been an impossible amount for a young labourer who, by saving and taking on a second job, could have accumulated the funds needed for purchase. And once in possession of this newfound mobility, it would of course have been quite natural for unattached working-class men (and possibly women), to dream of not only what, but who they might encounter. Cycling's growing accessibility does, in short, provide a highly common-sense explanation for the trends recognised by Perry.

The significance of these developments extends well beyond the thrill of romance, to the consequences: children born to parents without overlapping family trees. By allowing people to escape from isolated rural areas, the bicycle is widely recognised to have increased the diversity of gene pools with all the wider health benefits this brings. In the words of the biologist Steve Jones, 'there is little doubt that the most important event in recent human evolution was the invention of the bicycle'.[12]

Within the context of late nineteenth-century Britain, such a conclusion can only be seriously applied to those who came from the types of communities studied by Perry. When it came to potential marriage partners, well-connected members of the middle classes clearly had far greater choice than rural labourers and farm workers. Indeed, by the time one reaches those at the pinnacle of the social spectrum in first cousins Princess Maud of England and Prince Charles, cycling achieved quite the opposite effect, as early courting rides laid the foundations for their later inter-family marriage.

Nevertheless, even if it cannot claim to possess the same distinctive significance, cycling did unquestionably also aid the love lives of unattached city dwellers. But to understand the bicycle's impact in this area, we need to examine not just the newfound mobility which it offered. Instead this is a story of how this extra freedom was used to break free from a very different barrier to burgeoning romance: the chaperone.

'Handsome ineligibles'

Cycling's sudden acceptance as an activity suitable for members of both sexes in the mid-1890s placed it on a collision course with the long-standing Victorian practice of chaperonage. Although middle-class women were gradually enjoying greater freedom of choice in how they led their lives, those in earlier life stages still had various aspects of their existence overseen by older female relations. Even at Oxford, a place synonymous with women's increasing independence and access to education, female students were expected to be chaperoned when taking exams or attending lectures. One writer, looking back on the early years of the 1890s, commented:

> Very few girls rode in the park unattended by a groom, or drove in a hansom cab alone. They were not allowed to dine out, or pay country visits by themselves, and certainly no girl,

except those who lived in the sacred precincts of Belgravia (and never beyond), was allowed to take a walk without some sort of chaperon.[13]

Chaperonage was especially designed to regulate the relationships between unmarried men and women. Clearly, being accompanied by a wary aunt or mother while out in public did much to prohibit contact from hopeful male suitors. Activities which were approved for inter-sex mingling, such as tennis matches, boating trips, picnics and other outdoor outings, were highly regulated affairs, designed to ensure the coming together of socially viable couples. Once a young pair had passed through this initial stage and started courting, chaperones continued to oversee their relationship and ensure that nothing 'untoward' occurred between them. This is reflected in a poem titled 'Of a Chaperon' in the *Hampshire Telegraph and Sussex Chronicle*:

> I sing of her whom maidens scorn;
> But men must tolerate;
> A personage society-born,
> An antidote to fate.
> Love laughs at bolts and bars, they say,
> But Cupid is outdone
> By one who's ever in the way –
> I mean the chaperone.
>
> She catches every whisper low;
> She intercepts each sigh;
> Each tender look she seems to know,
> And naught escapes her eye.
> Her sight is keen, her hearing good,

> Her heart is like a stone;
> And no man ever understood
> The wicked chaperone.[14]

It was this tightly controlled world of courtship and romance that the bicycle did much to transform. For there was no other activity that so ideally lent itself to escaping the watchful eyes of parents and other interested parties than cycling away into the quiet of the country-side. 'The chief merit of the bicycle in the eyes of the young is that it dispenses with the chaperone,' commented one newspaper article, while Florence Harcourt Williamson likewise observed:

> The beginning of cycling was the end of the chaperon in England, and now women, even young girls, ride alone or attended only by some casual man friend for miles together through deserted country roads.[15]

Cycling certainly posed significant problems to the older women who had traditionally been responsible for watching over unmarried girls and their admirers. With few women of their age having had any reason to ride before, even before facing the challenge of keeping up with sprightly young charges, they had to pass through the painful process of gaining mastery over their machines. In her book *A Wheel within a Wheel*, Frances Willard, aged fifty-three, philosophically detailed the difficulties which came with learning to cycle at a later stage in life. Willard calculated it took her a total of three months to become fully confident atop 'this most mysterious animal', a process which saw her struggle to overcome many difficult and painful incidents along the way. For example:

One bright morning I bowled on down the Priory drive waving my hand to my most adventurous aide-de-camp, and calling out as I left her behind, 'Now you will see how nicely I can do it – watch!' when behold! that timid left foot turned traitor, and I came down solidly on my knee, and the knee on a pebble as relentless as prejudice and as opinionated as ignorance.[16]

Although Willard may have exhorted her readers to follow in her footsteps and master 'the most remarkable, ingenious and inspiring motor ever yet devised upon this planet', not all those her age reached the same conclusion. The assorted cuts and bruises which came with learning to ride acted as a powerful deterrent. One newspaper article, which interviewed various women as to the reasons why they had started cycling, included one mother who felt it her 'unpleasant duty' to take up the pastime alongside her daughters. With all their friends now going off on two wheels:

I had to let my girls learn to ride too; and as I cannot let them go alone, I have had to learn as well in my old days, although it is torture to me. Do you think I would be such a fool as to ride at my age if I were not positively obliged to do so?[17]

For the interviewee and other similarly minded mothers, 1896 brought the welcome news that a 'Chaperon Cyclists' Association' had been founded in London. Aimed at the large target market of mothers who did not want to spend their Saturday afternoons struggling to keep up with their daughters, the association was set up to provide a ready supply of wives, widows and unmarried ladies over the age of thirty who could go in their stead. All recruits to the association were required to be 'accomplished on wheels' and 'unimpeachably respectable', charging three shillings and sixpence (around £11 in today's money) for an hour's chaperoning, and ten shillings and sixpence (about £32) for a whole day.[18]

The fact that there is no further mention of the organisation after 1896, however, suggests it did not last long. Indeed, even by the standards of the time, the association was viewed as being comically out of date. As *Cycling*'s female columnist semi-seriously noted, 'a high-spirited, energetic girl, on the light side of twenty' could easily escape the attention of her overseer should a 'handsome ineligible' appear, while another newspaper commented:

> The cycling young lady is generally able to take care of herself without the aid of a chaperon, and we have an idea that wives, widows, and spinsters above a certain age are not very congenial company for her.[19]

Although it would be too much to suggest that cycling signalled an end to chaperonage, it did, in the words of David Rubinstein, deal the system a hearty blow.[20] For young women, cycling offered a ready and, from the mid-years of the 1890s, acceptable means of breaking free from the restrictions previously placed upon them. As such, the question now facing them was a simple one: how to best use this new freedom?

'Fatal forms of sentimentality'

For those with one eye on meeting prospective partners, weekend trips into the countryside certainly came full of possibility. As well as bringing you into regular contact with individuals from outside your social circle, cycling also had the welcome habit of manufacturing opportunities for casual encounters – none more so than the challenges arising from a punctured tyre. The labour and blackened hands required to mend an inner tube ensured this procedure was commonly viewed as 'unladylike', meaning it was a skill possessed by few female cyclists. As such, when faced with difficulty they were often required to rely on the help of passing male riders. One lady member of the Tottenham Cycle Club stated:

> I doubt very much if many ladies mend their own tyres. I confess I should have to look for a mere man in such a case, and would welcome his assistance, as I don't feel anxious to learn how to do it, having no fondness for pinching my finger.[21]

H.G. Wells developed such a situation as the motif for his short story, 'A Perfect Gentlemen on Wheels'. Published in 1896, it revolves around the meeting between the main character, Mr Compton, and Ethel, a female cyclist whom he meets by the side of the road struggling with a deflated tyre. At the point when Compton stops to offer her his assistance, Wells confides to his reader:

> Now this is the secret desire of all lone men who go down into the country on wheels. The proffered help, the charming talk, the idyllic incident! Who knows what delightful developments?[22]

That a certain amount of truth resided in his lyrical assertion can be seen in a piece in the gazette of the Stanley Cycling Club, which commented:

> We do not for a moment wish to assert that the cycle is… a cure for single blessedness, but, nevertheless, there are a few cases on record relating to both sexes. One or two prominent men of cycledom took an overdose of cycling some years ago, and caught a wife. Two or three girls we know of found 'friends in need' which later turned out to be friends indeed, for they married those girls.[23]

For those starting to move beyond their 'single blessedness' and otherwise engaged in courtship, cycling also came with many useful aspects. While flat tyres may have been less conducive to romance when riding out as a couple, otherwise unpleasant features of the pastime still offered opportunities to become more intimately acquainted. The common teaching method of an arm being placed around the waist of a struggling pupil, for instance, offered the rare prospect of unmarried middle-class men and women experiencing close physical contact with each other. An article in the *Aberdeen Weekly Journal* from 1896 described how 'Albyn Lane, the once secluded vernal bye-way sacred to whispering lovers', had now become 'the happy practising ground of female cyclists' who

> … any fine evening… may be seen acquiring the art with the aid of their male friends and advisors, and the comicalities of the situation – the ill-concealed flirtation – the agonising efforts to sit on the 'byke' – form a spectacle worth a visit.[24]

While acquiring mastery of a machine may have encouraged romantic entanglement, the real benefits of cycling arrived once couples were confident enough to ride away into the quiet and seclusion of the countryside. Away from Aberdeen's Albyn Lane, there was Love Lane in Lincoln which, according to local folklore, acquired its title from the cyclists who flocked there eager to find isolation on the city outskirts. ('Love Lane' has rather sadly since had its name changed to Milman Road, and been absorbed into the city itself.)

As well as privacy, the physical release afforded by cycling into the countryside also had the beneficial effect of relaxing and loosening up the behaviours of budding couples. 'There is something about a tandem, if it steers easily, which softens the iron-bound laws of custom' remarked an article in the *Leeds Evening Express*. According to a piece on 'Cycling Courtships' in *The Hub*:

> Cycling parties are got up, and the ride is still young when the cyclists tend to sort themselves into couples, each couple consisting usually of a lad and a lass. And soon the mental as well as the physical effect of the ride commences its influence. The exhilarating exercise gives a sense of well-being, and tends to lessen restraint and conventionality of manner.

According to the writer of the piece, 'rushing through wooded tracts and sunlit meadows' was conducive to 'fatal forms of sentimentality', ensuring that within very little time, 'Jack is speaking "sweet nothings" to Jill, who is too happy and light-hearted to snub or repel him'. The now familiar matchmaking device of flat tyres plays its part too:

> There are many punctures done on purpose, which necessitates a tête à tête walk home – for surely no gentleman would

allow a lady to walk home by herself – in the gloaming, or
nuts may be lost (or carried in the pocket); and the stars are
peeping before the weary, worn, and travel-stained couple
arrive home full of anathemas upon their misfortune, but in
reality, probably, if not engaged, often on the brink of an
engagement.[25]

We can wonder whether a solitary afternoon ride would always have
resulted in young men and women so speedily passing through the
courting process. Nevertheless, it is impossible to deny the profound
impact cycling had on middle-class courtship rituals. The mother
who described the 'torture' of learning to ride bemoaned the fact that
young men familiar with her daughters 'used to come to one's house;
now their bicycling excursions always prevent them from doing so, and
one is always hearing that Miss So-and-So is going with them'. After
initial resistance, she and her husband finally gave in to their daughters' requests for new machines after they had discovered that 'two of
my husband's nieces, who are not anything like so pretty as my three
girls, had got engaged whilst bicycling'.[26]

All of this suggests that, while loosening some conventions, the
bicycle's impact on pre-marital romance was broadly accepted as
benign, accelerating the matchmaking process while remaining within the bounds of propriety. The image of male and female cyclists
speeding down the path to matrimony provided a reassuringly familiar
view of relations between the sexes. Although a 'forward' new social
convention, courtships that took place on bicycles were still closely tied
into older traditions. It is telling that in fictional depictions of rationally dressed 'new woman' cyclists, romance was frequently the route to
curbing their adoption of more radical, new-fangled habits. Take, for
instance, a poem on 'The Lady Scorcher' in *Cycling*:

> I met her on the Ripley Road,
> Sitting her Humber like a queen,
> She wore the knickers à la mode,
> Whereby her dainty limbs were seen.
> A 'Rights of Woman' maid was she,
> And, of a truth, was fair to see.

After trudging its way through a couple more verses written in similar vein, the piece concludes:

> She rides no more the Ripley Road,
> Since Cupid came in view,
> Her song has changed from 'Woman's Rights'
> To 'A bicycle made for two'.[27]

Of course, approval of unmarried couples going off on unsupervised excursions together was also underpinned by an understanding that although 'sweet nothings' might be whispered, their relationships while out riding would remain platonic. It seems likely that such trust was well placed. After all, for all the seclusion and freedom offered by cycling, it was still a public activity typically enjoyed in the broad light of day. Parental approval depended on couples returning at a reasonable hour. It was unthinkable that those yet to be married might enjoy overnight hotel stays together. Though cycling may have loosened up behaviour and accelerated the courting process, in its socially approved forms it still fell a long way short of shaking the foundations of Victorian morality.

But perhaps this isn't the full story. We only need to travel back to the activities of clubmen detailed in the previous chapter to remember that significant divides can exist between approved, public representations of cycling, and what in fact occurred when people went out riding. The somewhat sanitised accounts that we can access through newspapers and other conventional publications only paint part of the picture. By returning to club gazettes, we can access far more revealing and lively accounts, told in the authentic voices of those living through the experience. What they reveal are tales of a very different nature. Not the relatively prim and proper rituals of courtship and marriage, but the archetypal adventures of men travelling away from home in search of erotic encounters.

'Julieting tendencies'

Interactions with women during weekend excursions certainly represented one aspect of club life that members were particularly keen to bring to the attention of their gazette editors. Within organisations whose runs and tours took place in exclusively male company, successful encounters with 'the fairer sex' represented a tried and tested way for clubmen to win the respect of their peers. This can be seen in a limerick published in the gazette of the Bristol B.T.C.:

> There's a sweet little girl at Fairford,
> By the Chappies round there she's adored,
> But without any doubt,
> I can knock 'em all out,
> Said our gay mashing member, who scored.[28]

As personified by the 'gay mashing member', when heading out on their weekend excursions clubmen eagerly pursued opportunities to pull off brief status-enhancing liaisons. Within the atmosphere fostered by cycling as an all-male group, members acted far more brazenly than if they had been out by themselves, or spending time with women from their home communities. 'Great excitement among lonely female riders on the road,' stated an article describing a Stanley Cycling Club run, while another piece in the club gazette described how, upon stopping at an inn on a club tour:

> We went, and made the brief acquaintance of a young lady who it had pleased providence to 'call at the bar'. I say brief, because we soon got into her bad books, and had to leave.[29]

Clearly, in viewing their encounters with 'young ladies' as opportunities to impress their fellow riders, clubmen were hardly inclined to treat those they met in a thoughtful and considerate manner. Descriptions of these meetings are written almost exclusively from the perspectives of members: writing at the start of the club run season in April 1896, one member of the Hull St. Andrews Cycle Club recalled:

> [T]he village feast of last year where each and all the crew imprinted a kiss on the lips of some rustic damsel and swore to be true to her evermore. But the following week they did the same thing again, and the damsels – God bless 'em – the damsels, well perhaps it would be best to draw the veil (alas, alas ye saints, what a record of broken vows and blighted affections is yours).[30]

It is difficult to take too seriously the image of bumptious young cyclists leaving a trail of broken hearts through the villages they visited on their rides. Indeed, rather than pining for their return it seems far more likely that the 'young damsels' described would have viewed the visiting clubmen in a similarly bemused fashion to Flora Thompson in *Lark Rise to Candleford*. As with the 'townsmen out for a lark' she portrays, members of the Hull St. Andrews Cycle Club also possessed a 'great sense of their own importance', which was greatly flattered by the idea that they were irresistible to the rustic women they encountered. Seducing and then abandoning the 'maidens' whom they met at the village feast strongly fed into the writer's assertion that:

> The cyclist is a merry fellow, an optimist of the best type, and given a good road, a good mount and jolly company he asks for nothing better. With song and anecdote the hours swim by, the cracking of jokes and the cracking of crowns are as pleasant to him as to the roystering [sic] old Cavaliers of Charles.[31]

The meetings with the opposite sex detailed in club gazettes are not the most reliable of sources. They provide just one side of the story, and there was ample scope for clubmen's pens to run away with them as they exaggerated and embellished the events described. Although the writer of the piece above may have envisioned club members as flamboyant and alluring cavaliers, one doubts that all they met during their rides shared in this impression.

Fortunately, there are articles which provide more honest accounts of what occurred during these brief encounters. Once you get past their bravado and posturing, there can be little doubting that clubmen found the prospect of finding romance on their rides genuinely exciting and alluring. This was especially true of long-distance tours, when they stayed overnight in inns and hotels. Combining generous amounts of cycling, alcohol and time away from the home, these events were liable to unlock the romantic sentiments of members no matter what their age. One younger Stanley Cycling Club member gave the startled account of how during one overnight stay,

> 'Betsy' fancied our Scotch, and after once or twice remarking 'Lor' lumme, but this is grand scotch', became confidential, and told me of his early loves, and so to bed. 'Betsy' beseeched me, by the memory of his youthful loves, not to snore, or he would be unable to sleep a wink all night; he wanted to dream.[32]

If Scotch whiskey helped those in later life dream once more about their youthful loves, younger members found much more immediate sources of stimulation. Their lodging places regularly brought them into close contact with women of a similar age working and staying there. The effects this could induce in unattached young cyclists are delightfully detailed in an article in the Bristol B.T.C. gazette. Written in characteristically extravagant style by 'The Unchained', the events described occurred after he and five of his junior clubmen had decided to stay on for an extra night during a club tour to Stratford-upon-Avon. They pitched up at a hotel where

> … our hostess was the happy mother of several lovely daughters, all in maiden meditation fancy free, and there were several previously unscotched hartz [unscorched hearts] beating beneath the buzzims of as many cyclists.

After an afternoon pleasantly spent 'boating on the Avon', members returned to their guesthouse to spend the evening. If 'The Unchained' was hoping to fill this time enjoying food, drink and lively conversation with his young charges, however, he was soon to find himself disappointed. Leaving them in the company of an 'American who knew how to spin 'em' in the hotel's smoke room, he briefly disappeared to write a letter for home. Upon returning, however, he faced a puzzle:

> Where were our merrie men; not in the smoke-room, not in the coffee-room, not in the bar. Methinks I will leave the house and find them. Aha! the mystery was solved. It was Romeo and Juliet all around. It must have been due to the atmosphere of Shakespeare's town.

Not only had his fellow members discovered the whereabouts of the landlady's daughters, but they had also hit on a quite novel means of trying to seduce them:

> Approaching one doorway I overheard, 'The brightness of thine eyes outshines the stars.' Can't pass that way; let's try another. Here 'twas 'Oh would I wear a glove upon that hand that I might kiss that cheek.' Cheek, indeed: what next. Next, 'Good night, good night, parting is such sweet sorrow,' &c. It is evidently no place for me. I will betake me to the back entrance and escape. Foiled again! A verandah was there, and it was the balcony scene over again – 'Oh swear not by the moon, the inconsistent moon.'

With this being 'too much Romeo-cum-Juliet' for the unfortunate narrator, he loudly proclaimed 'that young man will swear by or at anything marm!' and took his unwanted presence back to comfort of the smoke room and the society of the American traveller. None too pleased, however, at being abandoned for the 'Juliets' outside, he wrote:

I felt half-inclined to go and make love [a term for chaste, romantic courtship in this period] to the *Mater* of those maidens, marry her right away, and, out of revenge, to act the part of the irate parent when my moon-struck fellow-sighclists came to ask our consent. Prudence prevailed, however – I had previously witnessed some of them perform with the gloves at our gym.

Realising he was in no position to act the part of interfering family member, 'The Unchained' left his younger clubmen to the company of their newly acquired interests for the rest of their stay. Upon departing from Stratford, and 'not wishing to be *de trop* as a witness to the delightful lingerie of the partings of so many Romeos' and Juliets', he undertook a one-man tour of Shakespeare's town, and 'for some time had he to await the Romeos approach'. Finishing with the familiar trope of cyclists leaving behind their broken-hearted lovers, the piece ends:

Ah, me! The Juliets have doubtless ere now consoled themselves with the Immortal's words –

'Sigh no more, ladies – ladies sigh no more,
Men were deceivers ever;
One foot on sea and one on shore,
To one thing constant never.'[33]

This was not, however, the end of the story. For after the article appeared in the club gazette, word soon reached the partner of a member present that evening, and in the next edition 'The Unchained' detailed the fall-out:

One of the Romeos referred to in my scribbles anent Stratford, complains of the remarks therein on the Julieting tendencies of the tourists. Now, his only ownest Dorothy Dimple is aware he came with us to Stratford, and having read Club Tours, there r rukshuns in konsekwense [ructions in consequence].

Faced with this difficult situation, he offered this none-too-apologetic advice:

There have been two club tours to Stratford. If he was present at No. 1, the amorous bacilli were on the war path and infected the Romeos of the second tour. If he came on the second tour, put it down to No. 1. If he attended both – well, he had better own up and kiss his Dorothy into forgetfulness.[34]

Members of the Bristol B.T.C. were certainly not alone in experiencing an infection of 'amorous bacilli' after encountering women in the places they were staying. Containing much the same ingredients as the piece above, an article in the gazette of the Argus Bicycle Club detailed what had occurred after two members had gone on an Easter tour together. The writer, again taking on the role of disgruntled observer, described how on the first day of their trip he and 'Ern' had stopped at a country hostelry for food, which was served to them by 'the landlord's pretty daughter'. Painting a picture of his touring partner, it was at this point that he remarked:

Ern, it strikes me, knew more about this place than he cared to tell me. He, I may say, is a bit of a dandy when a girl is in the way, whether she be Lady Donowho [don't know who], or a domestic servant, but I never saw him anywhere so smitten as with this girl; he rang the bell times out of number for fresh things, impossible things mostly, but for the purpose of speaking to her, and she positively humoured him.

Rarely for club gazette pieces, the piece then features the voice of the female figure subject to Ern's attention, as at the end of the meal she asks:

'Anything else sir?'

'Oh yes, Mary,' said Ern, 'I'm awfully fond of the name Mary, there is just one thing I need to make this the nicest dinner I've ever had.'

'What is that, Sir?'

'Come and give me a kiss.'

'Oh how can I Sir,' demurely replied she, 'I haven't any, and so cannot give you what I cannot get myself.'

'Well, just wait a second and I'll lend you one, which you can pay me back by instalments and with interest.'

And she waited. I merely looking on, Ern was quite enough for her. No pretty maid for me.

Whether Ern's chat-up lines represent a step up from the Shakespearean efforts of the Bristol B.T.C. is difficult to say. But whatever judgement we reach, they clearly achieved their desired effect with Mary, meaning the writer was swiftly relegated to the role of awkward third wheel. He detailed how the afternoon passed in much the same way as lunch, with

Ern and Mary enjoying themselves, whilst I only attracted an occasional word from either. Finally, Ern decided not to take to the road again that day, as it looked cloudy and threatening, but I was determined to, and go I would, with or without him; and Mary, putting in a word, held him to his decision, so I went forth alone.

With his excursion not panning out as planned, things proceeded to only get worse for the ill-fated storyteller. Although agreeing to meet him the following morning, Ern, not unsurprisingly, failed to honour their agreement, leaving him to press on unaccompanied. Already decidedly disillusioned, the author's mood darkened still further after the threatening weather from the previous day manifested itself as a heavy snowstorm. The article ends with the writer describing how he eventually returned home:

> Upon calling round at Ern's home, I found his people study-
> ing a map of England to find out where the place was Ern
> was at, for he had telegraphed (although the telegraph office
> was at the railway station) the laconic message, 'Snowed up
> in paradise.'[35]

As with other accounts in club gazettes, we can only speculate as to what may have taken place between Ern and Mary. But taken together, the pieces provide us with one clear take-away message. Namely, the privileges male cyclists enjoyed compared to their female counterparts. Although unmarried women may well have ventured away from home communities in search of dalliances of a similar nature, to do so would have been far more controversial and at odds with social convention. While the image of female cyclists riding away from their chaperones is highly evocative, escaping the restrictions of middle-class courtship came far more easily to one side of the gender divide than the other.

When it comes to pushing against the limitations of restrictive social practices, however, there is one hitherto ignored aspect of late nineteenth-century romance also worth considering. For all the different types of male-female relationship that we have so far explored share the same essential quality of being, well, male-female. Clearly, not all would have sought this same dynamic when finding a partner to share in their tours and excursions. It is therefore well worth asking whether we can discern if the bicycle was a facilitator not just for hetero, but also homosexuality?

Playing on young hopefuls

When exploring late nineteenth-century homosexuality, it is important to note that in the eyes of contemporaries this was an almost exclusively masculine issue. At a time when female sexuality was a taboo subject, lesbianism fell outside the middle-class imagination as an unrecognised and incomprehensible activity. This is far from saying that women did not privately enjoy lesbian relationships, with Mary Benson, wife of the Archbishop of Canterbury, a case of one woman who enjoyed several female lovers. Society remained oblivious to such unions, however, which is symbolised by the story that they were due to be criminalised in 1885, but supposedly avoided such a fate after Queen Victoria declared them to be impossible.[36]

By contrast, this was a time when men's sexual urges were commonly recognised. Acquiring these had long been viewed as a part of growing up, it being acknowledged that adolescents would typically struggle to control the newfound desires that came with puberty. In his book *The Threshold of Manhood: A Young Man's Words to Young Men*, published in 1889, the clergyman W.J. Dawson reassured his readers:

There is a period in life when the desires of the flesh exercise immense influence and subtle power over the imagination. They seem to promise illimitable delight and inexhaustible pleasure. They sting the flesh with their violence, and send the blood boiling through the veins like a tide of fire… That is not an unusual experience. It is common to all of us in the heyday of youth and strength.[37]

Such sympathetic understanding was predicated on the assumption that this was a period in life which would soon pass. While it was a phase which could lead to the sowing of 'wild oats', experience dictated that young men would in time mature, pass through 'the threshold of manhood' and settle down with a partner. Though not a subject for polite conversation, chasing after the landlady's daughters while on tour was an activity that could be quietly accepted as a case of 'boys being boys'.

Homosexuality, however, was a different matter altogether. As well as meeting with far greater ignorance and misunderstanding, within a still-Christian society 'sodomy' was seen as being at odds with the sacred institution of marriage and family life. Although young men could be forgiven for a lack of self-control when it came to women, homosexuality had habitually been targeted as a 'debauched' activity, stemming from far greater deficiencies in willpower.

While continuing to inform attitudes, in the closing years of the nineteenth century these older Christian viewpoints were bolstered by even harsher medical arguments. Same-sex passion was increasingly presented as 'perverse', being described in the same terms as mental illness or psychological disturbance. In his book *Sexual Inversion*, published in 1897, the influential physician Havelock Ellis doctored case histories to support his claim that no case of homosexuality unconnected with the asylum or the prison had ever before been recorded.[38]

Such assertions, which presented gay men as a highly dangerous and threatening 'other', were also supported by wider scientific discourses. This was a time when Darwinian theories of evolution had made it common to speak about nations as 'races' engaged in a life or death struggle for survival. Homosexuality was frequently presented as both a cause and a symptom of 'degeneration', in which the 'British race' was seen not to be evolving, but rather decaying and moving backwards. As with other 'decadent' social practices associated with modern city life, such as frivolous consumption and sexual excess, it was seen as threatening a societal decline such as that experienced by the late Roman Empire.

These newfound anxieties concerning homosexual activity are reflected in a hardening of the judicial code. Further to older laws which prohibited 'indecency' in public, the Criminal Law Amendment Act of 1885 saw private male same-sex acts severely legislated against, with gay sex behind closed doors made a criminal offence. The law formed the centre of the Oscar Wilde trial ten years later, which centred around allegations that Wilde had slept with several young men, including Arthur Douglas, son of the 9th Marquess of Queensberry. The case aroused huge public interest and controversy. Wilde was found guilty of 'indecent behaviour with men', earning him the maximum penalty of two years' hard labour and effectively ending his career as an author and playwright.[39]

In such a context, making any public mention of homosexuality would be ruinous – so it is unsurprising that the topic is not openly discussed in any source material. Though the activities of the Bristol B.T.C. in Stratford may have landed members in hot water with their intendeds, such consequences were mild when compared to public shaming and a prison sentence. In the absence of documentation, the question of homosexual relations between cyclists is one which has a very simple answer: we have no way of knowing.

Recognising the toxic atmosphere which surrounds homosexuality in this period does, however, give a fresh impetus to previous discussions. For however much they pursued women when out on runs, it is hard not to be struck by the contrast between the massive hostility directed towards sexual male relationships, and the widespread practice of men building highly emotional relationships in clubs that barred female members.

Although we can easily relate to the bonds which existed between clubmen, by modern standards their relations had a striking degree of intensity. In addition to several long-distance tours a year, the weekend runs of institutions such as the Stanley Cycling Club were overnight affairs, beginning soon after lunch on a Saturday and not ending until Sunday afternoon. Away from cycling, members also enjoyed regular access to a clubhouse which remained open throughout the week and provided a focus for their activities during winter months. This essentially functioned as a gentleman's club, containing a well-stocked library, an even better-supplied bar and dining area, as well as several pool and billiard tables. Writing in January 1897, one member reported:

> Those delightful Saturday evenings at the Club House are now in full blast. Solos and suppers, cigars and chat, dominoes and drinks on the first floor, pipes and pool on the second, where the puncher and the punchee become either the one or the other in turn and by turns; for good fun and fellowship, commend me to this apartment.[40]

Despite being pitched as a place where clubmen could find relaxed jollification, the lengths of time they spent enjoying each other's company inevitably fostered relationships which extended far beyond the casually sociable. In another piece, it was detailed how upon revisiting the club house after a long absence, a former member 'at once fell on his neck and wept – à la Dreyfus – and the atmosphere of the library became humid with salt tears and Irish whiskey'.[41] Examples like this lead one to ask how clubs that facilitated such deep and expressive male relationships can be squared with the widespread homophobia that existed in the society at the time.

It is quite natural to wonder if the former was a consequence of the latter, as members sought outlets for otherwise suppressed same-sex desires. Such motivations may well have existed on an individual level. After all, large institutions with close to a hundred members would have surely been home to a handful of privately gay clubmen. Indeed, it is not beyond the realms of possibility that the privacy, relaxed nature and opportunities fostered by club life (it was a common feature of long-distance tours that members would share rooms and beds together) could have resulted in the realising of stifled passions. As with other bohemian and nonconformist groups from this period, institutions such as the Stanley Cycling Club may even have been prepared to acknowledge and turn a blind eye to such behaviour. Homoerotic innuendo can be found in the club gazette, with one article describing a club house game of pool:

> The sight of 'Cholly' King chollyking his cue preparatory to 'playing on' the young hopeful, is a study in family amenities, and the twinkle in Gussy Lumb's eye as he puts 'Dan'l' to bed for the third and last time, sets off a roar of laughter which entirely drowns the seductive proposal of the marker to 'star one, Sir'.[42]

Whether undercurrents of same-sex desire fed into some men's decisions to join cycling clubs, and whether these urges were ever realised,
can be treated as open questions. But at the same time, it is hardly
credible to suggest that shared homosexual passions were what drove
thousands of men to seek out the exclusive company of other male
cyclists. To explain the contradictions highlighted above, we are better
off looking at men's wider relationships with those whom these clubs
were designed to provide escape from in the first place: women.

New forms of comradeship

When exploring middle-class gender relations in the 1890s, it is hard
to feel a great deal of sympathy for men. As we have seen on various
occasions, 'the sterner sex' had it far easier when it came to enjoying
the various opportunities created by the bicycle. In nearly every aspect
of the pastime, from scorching to engaging in brief dalliances, widely
held notions of 'masculine' and 'feminine' behaviours afforded male
cyclists much greater freedoms than their female counterparts.

For all the privileges which they conferred on men, however, it
is important not to let this perspective override what was a more
complex picture. Although coming down heavily on the 'manly'
women who raced in rational dress, polarised understandings of
gender were also highly critical of men who made the reverse journey across their tightly drawn limits. In the case of homosexuality,
they further encouraged a perception that this was 'unnatural', being
completely at odds with the accepted norm of heterosexuality. A criticism frequently levelled at Oscar Wilde was his 'effeminacy', as he
brought into disrepute prevailing understandings of how men were
supposed to behave.[43]

If strict notions of men and women's different natures further explain the hostility directed towards homosexuality in this period, then they also do much to account for the popularity of institutions such as the Stanley Cycling Club. For the main environment where members might otherwise spend their free time – the home – was a space centred around understandings of sexual difference. Middle-class ideals of domesticity, hugely influential during the Victorian period, split the responsibilities of earning a living that would provide for a family, and running and overseeing the home, along a strict male-female axis.[44]

Building on older traditions, domestic life was presented as the harmonious working of men and women's different natures and capabilities. With husbands responsible for their family's material well-being, the home was seen as a space where they could leave behind work-related stress and worry, as they opened themselves up to the charms of domestic life and feminine care. 'Ah, what so refreshing, so soothing, so satisfying, as the placid joys of home!' rejoiced Reverend W. Jay in a piece on 'domestic happiness' written in 1849, as he pictured a 'man of trade' and asked:

> What reconciles him to the toil of business? What enables him to endure the fastidiousness and impertinence of customers? What rewards him for so many hours of tedious confinement? By and bye, in the season of intercourse he will behold the desire of his eyes and the children of his love, for whom he resigns his ease, and in their welfare and smiles he will find his recompense.[45]

The extent to which men shared in Jay's perspective and ideals did of course vary considerably. The prospect of returning home to tea on the table and time to be spent with loving wives and children was undoubtedly a source of sustenance for many. At the same time, cycling club gazettes reveal that many members viewed the duties which accompanied married life with a definite lack of enthusiasm.

This mindset was partly a consequence of the lost autonomy which came with settling down. Although recognised as the head of a household in a 'women's sphere' from which they were absent for long stretches of time, men often found themselves out on a limb when it came to domestic routines and rituals. The prospect of being cornered into activities which few associated with happy escape from 'the toil of business', such as afternoon tea and entertaining relatives, evidently filled many a clubman's heart with fear and dread. Further to being accused of selfishness when he looked to go out riding, one recently married Argus Bicycle Club member (who only dared sign the letter he sent in to the club gazette as 'The Personal Pronoun I') related how, upon telling his wife that he missed his former cycling friends and longed for society:

> The next afternoon I get it. My wife's mother, and antiquated aunt, come and take tea (enough to float a frigate), and try to cheer me – we do have a merry time – yet somehow it doesn't work; so now, when I am confidential and give vent to my feelings, I stay away and let my wife enjoy the company intended for me.[46]

If some counted the cost of domestic life through dull routines, then many others focussed on its financial outlays. With it being expected that middle-class men would serve as the sole breadwinner for their families, after entering matrimony their income soon needed to stretch to maintain a home, wife and children. By divesting their salary to support others, marriage required men to shoulder far greater economic responsibilities compared to when they were single. After exhorting his fellow clubmen 'not yet noosed' to 'revel in your freedom, laugh at your folly, thank your cycle for your escape from an unknown bondage, determine to remain single and retain your single blessedness', one Stanley Cycling Club member praised cycling as

> ... a less expensive luxury than marriage, you know exactly what your outlays will be before you start, the object of your affection has no costly tastes, no fickle whims and fancies; you need not reclothe it forty times a year, nor make great fuss about its suitable accommodation in magnificent apartments.[47]

Whether real or imagined, such contrasts ensured that for both single and married men there was an obvious attraction to spending their weekends with male cycling club mates. Doing so carried none of the same restrictions associated with the home and wider domestic life. Free to indulge behaviour which would have been deemed inappropriate when in the company of women, their interactions were also defined by what they had in common: a shared love of cycling and a mutual desire to enjoy their leisure time. The writer of the *Cyclists' Touring Club Gazette* article on 'The Elderly Young Man' from the previous chapter offered the further character assessment:

> A cycling holiday is really a holiday for him. I have seen him on his (nominally) real holiday with his wife and family, and a more miserable sight I do not want to behold; his mind evidently overburdened with thoughts of lodgings, cabs, lost luggage, and family cares. The few days he gets at Easter and Whitsuntide with us are the only real days of life he has.[48]

Cycling clubs were certainly far from being the only all-male environments to which middle-class men flocked in this period. Not only did the later years of the nineteenth century see a rapid growth in sporting associations, whether golf, cricket, rugby, football or athletic, this period was also the heyday of the gentleman's club, celebrated for allowing men to 'step outside the circle of domestic cares, and discuss masculine topics in a masculine atmosphere'. At a time when the average age a middle-class man married had crept up to thirty, the 1880s and 1890s are recognised as decades in which they were less enamoured with domesticity compared to earlier in the century, and increasingly sought out male company in their free time.[49]

However, when it came to middle-class gender relations, the bicycle's impact was without doubt most keenly felt in how it brought men and women together. While all-male institutions were an established feature of the pastime, by the later years of the 1890s, as we have seen, they were being subsumed by the mixed-sex organisations that promoted 'frank, friendly comradeship amongst men and women'. You also only need return to the renowned clubman A.J. Wilson, who found 'no form of cycling so quietly enjoyable' as going out with his wife on their sociable, to remember that while cycling may have allowed some to break free from married life, for many others it enhanced and enriched it.

Indeed, the sociable and romantic relationships which people found through cycling were closely entwined. Elizabeth Pennell's descriptions of her tours with Joseph, full of humour, teasing and light-hearted affection, could in many instances have been lifted straight from the pages of a club gazette. The pleasure and excitement which came from building these multi-faceted relationships are wonderfully brought to life in H.G. Wells's *The Wheels of Chance*, in which, after the first morning of cycling alongside his budding love interest Jessie, Hoopdriver meditates on

21 The Anfield Bicycle Club photographed whilst on tour in North Wales, including one member wearing an oversized watch and another in a sporran (1880s).

22 Fashionably dressed and smiling couple photographed with their bicycles (date unspecified).

… the intoxicating thought of riding beside Her all to-day, all to-morrow, perhaps for other days after that. Of talking to her familiarly, being brother of all her slender strength and freshness, of having a golden, real and wonderful time beyond all his imaginings. His old familiar fancyings gave place to anticipations as impalpable and fluctuating and beautiful as the sunset of a summer day.[50]

Enabling men and women to share in the excitement of journeying away from domestic environments, cycling did undoubtedly bring the sexes together 'on more equal terms more completely than any previous sport or pastime'. It spoke to a new vision of middle-class gender relations, with a greater emphasis on what united as opposed to what separated men and women. And crucially, the benefits of these less restrictive understandings were felt on both sides of the gender divide. It is telling that the Stanley Cycling clubman so vocal in his opposition to the costs and restrictions of marriage wished for himself a 'cyclist wife', as he advised his fellow members:

Pick out a rider who will be your accompanist in cycling excursions, and you will both be happy for ever.[51]

5

Travel, Adventure and British Tourists

By the turn of the twentieth century, the country lanes which surrounded Worcestershire's Malvern Hills had garnered a well-deserved reputation as a happy hunting ground for the touring cyclist. With their breath-taking views and splendid road surfaces, they were an ideal destination for those eager to enjoy the delights of the British countryside. According to one *Cycling* correspondent who visited the area:

> The scene comes as a revelation – for miles and miles the eye wanders across the peaceful vale till it reaches the Malvern Hills, and it is said 12 counties are visible from this scenic standpoint – the Prospect. Right below... nestles Broadway, which village certainly took pride of place in our wheel wanderings. No desecrating railway fills this rural Elysium with ruthless devastators, and its beauty brooks no comparison, for comparisons are odious.[1]

Not surprisingly, other cyclists were able to express the beauty they experienced when travelling through the vicinity. Nowhere was this more surely exemplified than in one of the region's most famous native

23 Photograph of Elgar proudly holding 'Mr Phoebus', his 1903 Royal Sunbeam.

sons. Proud owner of a state of the art *Royal Sunbeam* bicycle, he was well known for riding along its lanes and by-ways on one of the day's most high-end models. Dressed in a smart knickerbocker suit, topped off with leather gloves and a bowler hat, he was a figure certainly likely to catch your attention if you chanced upon him during your excursions.

In the event of such an encounter it is unlikely, however, that he would have reciprocated your interest. This was not, as his bicycle and attire may have suggested, a consequence of social snobbery. For Edward Elgar had discovered that there were few better ways of finding inspiration for his compositions than immersing himself in bicycle rides around his native countryside.

In the early years of the new century, Elgar was enamoured by the bicycle. He ordered his *Royal Sunbeam* in 1900, and upon its arrival bestowed on it the nickname 'Mr Phoebus' after the Greek god of the sun. The rhythms of pedalling his machine, the sights and sounds of the surrounding countryside, and the mental stimulation he found through cycling came together to inspire his creativity at a time when he was producing much of the music that would affirm his place as one of Britain's all-time leading composers.[2] The climax of his work *The Apostles* is well documented to have emerged during his cycling expeditions, while his cycling companion Rose Burley wrote:

> Much of Edward's music is closely connected with the places we visited for, as we rode, he would often become silent and I knew that some new melody or, more probably, some new piece of orchestral texture, had occurred to him.[3]

A quite unique record, Elgar's music stands alongside countless other sources that bring to life the almost mystical experiences cycling created for late nineteenth-century men and women. These had, of course, been taking place well before the invention of the safety bicycle and pneumatic tyre. Ordinaries had offered even more spectacular views of the surrounding countryside as, perched high up on their machines, cyclists had been able to peer above hedgerows and better take in the scenery around them. And even cumbersome tricycles could induce soul-lifting effects on the men and women who rode them. Recalling a journey through southern France when she and her husband were blown along by a tailwind, Elizabeth Pennell wrote:

> We rode so fast, we only knew we were flying through this beautiful green world. The clear air and cold wind gave us new life. We must keep going on and on. Rest seemed an evil to be shunned. For that afternoon at least we agreed with Mr. Tristram Shandy, that so much of motion was so much of life and so much of joy:– and that to stand still or go on slowly is death and the devil. We said little, and I, for my part, thought less.

After rushing along for a while longer, Joseph eventually gave vent to feelings like Elizabeth's, as she recorded his joyful cry:

> 'Hang blue china and the eighteenth century, Theocritus and Giotto and Villon, and all the whole lot! A ride like this beats them all hollow!' he broke out, and I plainly saw that his thoughts had been more definite than mine.[4]

Crucial for enabling greater numbers to experience these pleasures, however, was the development of pneumatic-tyred safeties. Not only did they have the long-term effect of lowering bicycle prices, but the newfound practicality and comfort they offered made it far easier to taste cycling's accompanying delights. Following these two inventions, there was an outpouring of accounts, ranging from Elgar's music to books and newspaper articles, expressing emotions much like those uttered by the Pennells. H.G. Wells thus described the first day of Hoopdriver's cycling tour in *The Wheels of Chance*:

> A sudden, a wonderful gratitude, possessed him. The Glory of the Holidays had resumed its sway with a sudden accession of splendour. At the crest of the hill he put his feet upon the footrests, and now riding moderately straight, went, with a palpitating brake, down that excellent descent. A new delight was in his eyes, quite over and above the pleasure of rushing through the keen, sweet, morning air. He reached out his thumb and twanged his bell out of sheer happiness.[5]

If gently parading around a park in a fashionable new outfit was one way of enjoying the pastime, then this was quite another. After the hugely popular worlds of fashion and racing, and the sociable and romantic interactions which took place between fellow cyclists, we have now arrived at cycling at its most basic and elemental. That is, the relationships between people and their machines, and how ownership of a bicycle enabled them to travel, explore and experience a world that had previously been out of reach.

But while they are a central part of this book's final chapter, cycling's accompanying joys and pleasures are far from its only aspect. For the spectacle of the huge numbers who had suddenly taken up the pastime invariably generated a much broader social response. A new social phenomenon, it sparked a whole range of fears and controversies extending well beyond the fact that many of these new enthusiasts happened to be women.

Moreover, even when we look at individual cyclists, upon closer inspection it is apparent that cycle riding has never been solely defined by gratification and enjoyment. After all, no amount of improvement in machine design could eradicate steep hills, headwinds, rainy days and the many other less appealing aspects of riding a bicycle. And although they were much more risk-free than ordinaries, people quickly discovered that the nickname given to safeties was not without a certain irony.

'Headlong tearing toboggans'

Just a few years before Elgar found mental stimulation through his bicycle rides, another renowned intellectual developed a very different relationship with his machine. Future winner of both a Nobel Prize and an Academy Award for his playwriting and wider social commentary, George Bernard Shaw was one of countless others who learnt to cycle during the craze of the mid-1890s. Possessing a highly distinctive riding style, he underwent experiences awheel which show that there was far more to cycling than 'the pleasure of rushing through the keen, sweet, morning air'. As his biographer Michael Holroyd puts it:

> For someone physically timid, Shaw's experiments by bicycle were extraordinary. He would raise his feet to the handlebars and simply *toboggan* down the steep places. Many of his falls, from which he would prance away shouting, 'I am not hurt', with black eyes, violet lips and a red face, acted as trials for his optimism.[6]

Shaw's assertion, four years after learning to ride, that 'If I had taken to the ring I should, on the whole, have suffered less than I have, physically', speaks of the painful consequences which often followed purchasing a machine.[7] As we saw previously with Frances Willard, the initial process of learning to ride a bicycle could be an excruciating rite of passage, involving multiple falls and accidents. Shaw wrote to a friend in 1895:

> At Beachy Head I have been trying to learn the bicycle; and after a desperate struggle, renewed on two successive days, I will do twenty yards and a destructive fall against any profes- sional in England. My God, the stiffness, the blisters, the bruises, the pains in every twisted muscle, the crashes against the chalk road I have endured – and at my age too. But I shall come like gold from the furnace: I will not be beaten by that hellish machine.[8]

Even after learning to ride, Shaw remained decidedly accident prone. The fact that, as with most others new to the pastime, he not only had no previous cycling experience, but also none of managing a machine along public roads, ensured he continually courted danger and mishap. In another letter he recounted that during a ride along Pall Mall he had come face-to-face with a spooked horse pulling a Great Western Railway van. With it quickly bearing down on him, he proceeded to undertake one of his 'experiments by bicycle', as rather than getting out of the way,

> I went ahead gallantly, and hit the horse fair and square on the breastbone with my front tyre, fully believing that the most impetuous railway van must go down before the onslaught of

Bernard Shaw. But it didn't. I hit the dust like the Templar before the lance of Ivanhoe; and though I managed to roll over and spring upright with an acrobatic bound just clear of the wheels, my bike came out a mangled, shrieking corpse. It was rather exciting for a sedentary literary man like myself.[9]

Aside from horse-drawn railway vans, there were several others who played supporting roles in Shaw's cycling accidents. Most notable of these was one of a select few individuals who could rival his formidable intellect. Fellow future winner of a Nobel Prize, Bertrand Russell would go on to achieve widespread fame as a mathematician and philosopher, most famously for his best-selling *A History of Western Philosophy*. Setting out on a holiday ride to Tintern Abbey in Monmouthshire, the pair were accompanied by Shaw's long-time friend Sidney Webb, fellow co-founder of the London School of Economics and a leading thinker in his own right.

Given the collective brain power of the trio, one might imagine that negotiating a short-distance bicycle ride would not have posed too many difficulties. With their minds preoccupied with other things, however, they displayed a silliness characteristic of intellectuals venturing out into 'the real world'. Shaw recounted the affair to a friend afterwards:

On Thursday afternoon, on the road from Trelleck to Chepstow, we three rode on our bicycles down a steep hill on our way to Tintern Abbey. Russell is rather absent-minded, as he is pre-occupied at present with a work on non-Euclidian space. He suddenly woke up from a fit of mathematical absorption, and jumped off his machine to read a signpost.[10]

A catastrophic development, this placed Russell square in the path of an onrushing Bernard Shaw, who was otherwise enjoying a 'headlong tearing toboggan down the hill'. With no time to get out of the way a spectacular collision resulted, in which Shaw was catapulted off his machine and

> ... flew through the air for several yards, and then smote the earth like a thunderbolt, literally hip and thigh – also shoulder, very hard and wrists. 'All right' I shouted (as if there were any hurry about it now), 'I am not hurt' and bounded up, pulling myself all together instinctively... I picked up my bike and trundled it up the hill. At the top, I felt sick, and the hills and clouds and farmhouses began to tumble about drunkenly.[11]

Remarkably, after a ten-minute spell sitting by the roadside, Shaw got up and continued his journey on a bicycle which was still in good working order. Unfortunately, the same was not true of Russell. Not only was his machine broken beyond repair, but he also faced the rather more delicate difficulty of finding that his knickerbocker trousers were now 'demolished' (this was a part of the tale Shaw told with scarcely contained relish). Russell later recalled how he was forced to return home by train, thinking that with Shaw pressing on they would have no further encounters that day. However,

> It was a very slow train and at every station Shaw with his bicycle appeared on the platform, put his head into the carriage and jeered. I suspect that he regarded the whole incident as proof of the virtues of vegetarianism.[12]

Even by his own high standards, the scene Shaw played out with Russell was exceptional. For its sheer readability and slapstick humour, it stands at the pinnacle not only of his various misadventures awheel, but of the wider world of late nineteenth-century cycling accidents. This was no easy achievement, for by the mid-1890s more people than ever before were experiencing calamities on their bicycles. And even if those involved were not famous thinkers, such incidents still enjoyed a generous amount of newspaper coverage.

Bicycle faces

During the time of the craze, cycling accidents became established as a recognisable genre of press reporting. Regularly published as short pieces under headings such as 'Cycling Accident to a Dundee Man', these articles provided graphic accounts of the fates which had befallen riders caught up in collisions. A typical example can be found in the *Yorkshire Herald*, which described how, when cycling down a hill and faced with an oncoming cab, a young 'Miss Ada Seale' ran onto the curb of the pavement:

> With the force of the machine striking violently against the curbstone, Miss Seale was violently thrown into Mr. Epworth's shop window, a large pane of heavy plate glass being smashed. It was seen by the amount of blood which fell on to the pavement, that the young lady had been much cut and injured.[13]

Aside from their gory detail, what is especially noticeable about such pieces is that they were almost non-existent in previous decades. If by publishing grisly 'Cycling Accident' articles newspapers were hoping to shock and, on a certain level, entertain their readership, you would expect them to have been far more common in the days of ordinaries. As well as being a common feature of the pastime, the spectacular nature of falling off a high-wheeled machine would surely have generated excellent material for accounts like the one above.

What happened during the 'craze' was not just that more people were involved in cycling accidents. Newspaper editors also became aware that their readerships were suddenly much more attentive to them. Within the wider population, people were asking questions typical of the emergence of a widely popular new social activity. Is it safe? What are the risks of getting involved? Should I let my children participate in it? In an environment primed for sensationalism and fear-mongering the press was, as might be expected, only too happy to oblige.

Those who operated outside the world of daily newspapers were certainly quick to call them out on overstating the risks of cycling. 'More pedestrians than cyclists are killed in city streets in the course of a year: yet nobody contributes long articles about the "terrors of walking". This eternal prating about the "dangers of cycling" is so very foolish. It makes us quite tired,' complained a *Cycling* editorial in 1898.[14] In more comical vein, an 1897 article in *Punch* titled 'Wheel Wictims!' detailed cycling 'accidents' which it claimed came from the pages of the *St. James Gazette*, a London evening newspaper. The piece began by stating:

> The long and terrible list of bicycling accidents, which (at this time of year) we publish daily, still continues to grow. The latest batch is even more alarming than usual, and proves conclusively that no one with the smallest respect for their safety should ever be induced to ride a bicycle… we would

recommend bathing in the whirlpools of Niagara as, on the whole, a less dangerous recreation.

It then went on to detail some of these terrifying 'bicycling accidents'. This included a worrying case from the 'highland village of Tittledrummie' where a 'youth of fifteen' had, after falling off his machine, scratched his hand on a gooseberry bush. Although he was

> ... still alive at present, it is highly probable that he will develop symptoms of blood-poisoning in consequence of his misadventure, when tetanus will certainly supervene, and the fatal bicycle will have brought one more victim to a premature death.[15]

Evidence of the sensationalised climate surrounding cycling at the time of the 'craze' can be found in other areas. Away from newspaper obsessions with cycling accidents, there were widespread concerns about bicycle-related medical conditions, which leading medical men played no small part in promoting. These included the 'vibratory habit', defined by eminent physician Sir Benjamin Richardson as 'a kind of intoxication for movement, an over-desire for rapidity of motion, an impression that one must be on a bicycle and riding for dear life at all times and in all places'.[16] And if from our modern perspective the 'vibratory habit' reads as a ridiculous medical diagnosis, then spare a thought for the much-discussed 'bicycle face'. Writing in *Pearson's Weekly*, C.A. Pearson, later founder of the *Daily Express*, described this as resulting from

> ... the constant anxiety, the everlasting looking ahead, the strain on a nervous disposition which imparts a hard, set look

to the face, and gives a haggard, anxious expression to the eyes which is quite painful to observe.[17]

Alongside the 'bicycle face' were numerous other conditions said to result from cycling, ranging from the bicycle back (caused by excessive hunching over the handlebars), the bicycle hand (caused by excessive gripping of the handlebars) to the bicycle heart (caused by generally excessive cycling). Although in theory these disorders could afflict anyone, they were most often discussed in relation to female riders. With medical convention holding that women possessed more delicate and fragile constitutions than men, they were seen to be most susceptible to ailments brought about by a 'strained nervous disposition' and physical exertion. On the bicycle face, Pearson stated:

> It is noticeable especially with women who ride in London among the traffic – where no woman ought to ride by the way... The matter reduces itself to this, that if women wish to retain their good looks, they must give up the foolhardy practice of riding in crowded thoroughfares.[18]

For the most part, the cycling press gave such claims the same treatment as over-the-top fears about the dangers of cycling, dismissing them as exaggerated speculation that stemmed from editorial need for copy. Writing about Pearson and the bicycle face, *Cycling* pointed out:

> We know riders of both sexes who have ridden for lengthy periods... and the only alteration we have ever noted in the countenances of any one of them is that the complexion has invariably been improved.[19]

By emphasising that cycling was not only far less dangerous than commonly supposed, but could also have a beneficial impact on people, *Cycling* and other publications looked to pour as much cold water as they could on popular panics. They promoted an alternative medical view of the pastime, which emphasised its potential for improving health and curing long-standing ailments. Writing in *Cycling*, one medical man testified to the various conditions he had seen treated by this 'prince of pastimes', which included 'general health improved' (thirty-eight cases), 'indigestion cured or relieved' (eighteen cases), 'improvement in spirits' (sixteen cases) and 'pimples cured' (one case).[20]

But even the most supportive medical arguments would nearly always qualify their praise of cycling with one critical exception. For it was widely held that the health benefits detailed above would only occur if people took their exercise 'in moderation'. Those who cycled at anything more than a comfortable pace were seen as risking long-term and even fatal consequences. In a widely read pamphlet from 1896, George Herschell, M.D. warned that an 'immoderate use of cycling' was likely to cause heart disease, as he cautioned:

> At a moment when old and young and middle-aged are alike resorting to this form of exercise, recreation and locomotion, a word of warning may not be out of place. Without it, there is some fear that in many cases the craze of the hour may develop into the injury of a lifetime.[21]

Fleshing out Herschell's views on 'cycling as a cause of heart disease', another expert wrote:

> At first the rider who has been overdoing it soon recovers himself after dismounting, but the attacks become in time

more prolonged and increasingly severe: the muscle of the heart, gaining strength by repeated exercise, works with alarming force, sufficiently so in one case I know of, to actually, with each throb, shake the patient's bed. In time, too, the heart begins to ache, not acutely, but just a dull, gnawing pain, that becomes most wearisome.[22]

Although the author may have dismissed the bicycle face and similar conditions as 'twaddle', medical warnings about cyclists 'overdoing it' clearly possessed similarities with these more speculative ailments. Not only guilty of a certain amount of scaremongering, they were likewise heavily focussed on female riders. National Cyclists' Union vice-president Dr. Edward Turner used accepted medical arguments to support his case against female racing, stating:

In women the nervous system is generally more delicate and more highly strung than that possessed by the sterner sex, and so the extra work and pressure thrown upon it when training for and competing in contests of speed and endurance, is more likely to cause a breakdown.[23]

Turner went on to warn that the novice rider who cycled too far too soon 'may inflict injuries which endure for years, or in the best case leave her prostrate for some days suffering from fatigue fever'. Giving the example of one female rider who had, according to newspaper reports, been forced to abandon a race at the third mile 'owing to a severe attack of hysterics', he used this case study to claim:

Until, therefore, a woman can change her sex with her garments, she had better be content with non-competitive sport, and with the bounteous reward in health and energy that wisely regulated exercise will ensure her, and not descend into the arena or follow the example of the noble dames in the time of Rome's decadence.[24]

At a time when female involvement in cycling was subject to intense scrutiny, gloomy 'expert' prognostications like Turner's had a strong influence on wider discussions about how women could acceptably participate in the pastime. Although there was widespread acceptance of the benefits which came with cycling in 'moderation', contemporary medical opinion also supported arguments for limitations to be placed on how and when women cycled. While the cycling press fought back against press scaremongering about the dangers of cycling, they were highly sensitive to the controversies around female cycling and still supported medical views which prohibited women from cycling in a manner deemed to be overly 'masculine'.[25]

But although largely sheltered from the storms and anxieties which surrounded cycling during the 1890s, men were not completely removed from them. While doctors may have had less to say about the risks of men riding bicycles, as we saw with Bernard Shaw they were still just as likely be involved in widely reported cycling accidents. And when away from the urban environments where these mishaps most commonly occurred, they were also especially likely to feel the force of a body that started to take a much greater interest in the activities of cyclists in this period: the police.

'Bumptious bobbies'

In the many 'Cycling Accident' pieces published during the craze, cyclists often found themselves cast in a very different role to that of tragic victim. They also featured as villains who recklessly ran down innocent pedestrians before speeding away from the scene of their crime. A standard example of one of these accounts is a short description that appeared in the *Aberdeen Weekly Journal* which, looking south of the border, reported:

> The other night a crippled lad named George Devine (9), residing in Bradford... was knocked down by a cyclist, whose name is unknown. The lad's right leg was fractured. He states that the cyclist fell on him, and immediately on hearing him scream, jumped up and rode off.[26]

Articles such as this can be placed within the much broader backlash against 'scorching' riders. While the injuries that could be acquired when cycling attracted widespread interest, their relevance was mostly limited to those who had or were thinking about taking up the pastime. In contrast, the risk of being run down by a swiftly moving 'bloodthirsty desperado' was something which, in theory at least, could happen to anybody. As such, it provoked a response that extended well beyond the pages of newspapers, into the laws that governed cycling in this period.

Many of these were laid down in the Local Government Act of 1888. This ensured that, unlike other road users, cyclists had to use lamps to make themselves visible after sunset. Those caught without a lit lamp after the designated 'lighting up time' (published weekly in newspapers and the cycling press), were liable to pay a heavy fine. The act also stated that when overtaking, cyclists must always sound 'a bell or whistle, or otherwise, giving audible and sufficient warning of the approach of the carriage'. And although no formal speed limit was applied, as we saw previously 'scorchers' could still be prosecuted for 'furious riding', defined as being 'riding and driving furiously, or so endangering the life or limbs of any person, or to the common danger of passengers in the thoroughfare'.[27]

Within the cycling community there was a certain amount of resentment towards these restrictions. The fact that large horse-drawn vehicles avoided the lighting-up law was a particular source of angst and unrest. Not only did this create feelings of discrimination, it also ensured they were more likely to hit cyclists undertaking night time rides. 'If *all* were compelled to carry lights, the danger of collision would be greatly reduced' wrote one *Cycling* correspondent, describing how a 'fast-trotting cob harnessed to a light dogcart, without lights' had recently injured him and damaged his machine. The unfortunate victim wrote in to condemn both the law and the vehicle's driver, who headed off after 'the usual argument' ended with him 'refusing to give his name and threatening violence'.[28]

By far the most common cause of complaint, however, was the way the 'lighting up' law and other regulations were enforced. As concerns about 'scorchers' became more common, so the police became increasingly known for their hostility and aggression towards cyclists. Although a particularly intense attention was given to clubs' organised road races, anyone pedalling along popular routes could expect to receive close attention from officers. One club captain wrote wistfully in 1895:

> The cyclist of the present day sighs for the happy times of yore, when he could go where he listed on the highways and bye-ways of merrie England without the slightest fear of being hauled like a malefactor before the beak... Now, however, it is

with some misgivings that he leaves his home for his customary spin, for he is quite uncertain whether it may or may not be his fate to be arrested by some overzealous policeman and carried as a prisoner of war to the nearest lock-up.[29]

Policemen certainly displayed a quite remarkable enthusiasm for enforcing 'furious riding' regulations. Rather than waiting for riders to come to them, a commonly used ruse by officers was to position themselves at the bottom of hills, as they sought to catch out cyclists who were otherwise enjoying a momentary 'scorch'. One Tottenham Cycle Club member described how during a night-time ride he had descended a hill at ten miles an hour, and

> ... noticing something black at the foot rang the bell. Immediately the something in black began to wave its arms and cape taking up the whole of the roadway, and yelling 'Hi, go steady will yer. No scorching on this road or I'll — soon have yer off.'[30]

The disillusioned writer ended the piece by warning his fellow members against 'furiously loitering' in the nearby area. Unfortunately for him and many others, 'furious riding' was a law which came with plenty of grey areas that were open to interpretation. Although anything over 14 mph was generally classed as an excessively fast speed, in an age before speed guns it fell to an officer's judgement to determine how fast cyclists were moving. Not all agreed with the verdicts they reached. 'The large majority of police are very poor judges of pace,' lamented *Cycling* after numerous letters from readers complained they had been hauled up while travelling at a leisurely pace that endangered the 'life or limbs' of nobody.[31] One correspondent put forward this theory:

The real scorcher the police can't catch. The racing man 'out for a blind on the road,' flits by in safety. What happens? Someone must be captured. So it's 'name and address, please' for the very next man or woman passing who does but look absolutely incapable of going fast.[32]

In fairness to the police, hauling up speeding riders was no easy task. 'The elderly bobby, who's stuffy and cobby, ain't got arf a chance with a scorcher on wheels,' cheerily proclaimed the protagonist of *Punch*'s 'Song of the Scorcher'.[33] Although many police units possessed bicycles, from a stationary starting position it was near impossible to catch up fast-moving cyclists who ignored their instructions to stop. While some solved this issue by pulling in the next person to pedal past, others used their initiative to come up with ways in which they might 'capture a scorcher'. And though newspapers' suggestions to use lassos and enlarged butterfly nets were ignored, the methods they employed still displayed a generous amount of outside-the-box thinking.

One approach was to position a plain-clothed officer up the road from their colleagues, who then used a whistle to alert them of approaching furious riders. This gave them time to get in position for an early version of the police 'road block'. An article in the *Leeds Evening Express* revealed how local forces were using an officer on horseback to stand in the middle of the road, leaving 'scorchers' little choice but to slow down or risk a Bernard Shaw style collision. Police capes were also used to a similar end, as officers buckled their cloaks together and held them across the highway after being alerted to the presence of a speeding rider.[34]

If tactics such as this sound potentially dangerous, they were still far safer than the methods employed by officers who acted alone. Unable to bar a cyclist's path, they instead relied on the element of surprise, springing on unsuspecting riders from concealed hiding places. Clearly keen to keep its readers up to date with the latest police strategies, the *Leeds Evening Express* reported one case in which a lampless cyclist had been travelling along a quiet rural road at night, when he suddenly encountered an officer who 'in his zeal for due enforcement of the lighting up law, gave no warning whatsoever, but rushed suddenly out of a gap in the hedge and laid violent hands on the rider'.[35] In similar vein, one *Cycling* correspondent wrote:

> Two friends and myself were returning from Ripley, last night, on a tandem and single. When we reached Ditton Marsh a plain clothes police officer (No. 65) rushed into the road, at the same time catching hold of the handlebars of the tandem, throwing us off. He went through this idiotic and dangerous performance because, in his opinion, our light was not giving 'sufficient' light. We said he was exceeding his duty. '65' thought otherwise.[36]

To add insult to injury, in cases such as this it was common for those involved to receive a heavy fine. Cyclists deemed to have been 'furiously riding' or travelling without a light could pay as much as £2 plus costs, an amount which would have been more than the weekly wage of less well-off enthusiasts. Given that in many instances those accused had not done anything wrong (when it came to judging contrasting testimonies, magistrates were far more likely to side with a police officer), it is understandable that *Cycling* put forward a view of cyclists as being subject to police 'tyranny' and 'persecution'.[37]

No matter where cyclists took themselves during the 1890s, anxieties and controversies inevitably followed. Indeed, when reading accounts of cycling accidents, the potentially fatal consequences of 'immoderate' cycling, and the aggressive activities of police officers, it can feel strange that anyone from this period felt brave enough to leave the safety of home and venture out on their wheels. You only need to return to the beginning of this chapter, however, to realise that no matter what these apparent dangers, they paled in comparison to the pleasures that could be found when pedalling out into the countryside.

Hobbies and peregrinations

> Partaker in my happiest mood,
> Companion of my solitude
> Refuge when gloomy thoughts intrude,
> My bicycle to you I sing!
> With you no cares my brain oppress,
> I laugh at fortune's fickleness;
> No other sports your charm possess,
> Nor match for me the joy you bring.[38]

For those of us who have tasted the various pleasures of cycling, it is incredibly easy to relate to the joyous odes that late nineteenth-century men and women wrote to their machines. The delight of rushing along country roads and taking in the surrounding air, scenery and sunshine fully transcends the distance in time that separates us. While we are removed from the author of the piece above by over a hundred years, his lyrics could, under a more poetically minded editor at least, easily find their way into a cycling magazine of today.

Having said this, however, it was also the case that those living through this period viewed the bicycle through lenses unique to their moment in history. Given that most had never engaged in a comparable activity, the sensation of flying along on their machines was especially exciting and novel. Furthermore, after a century of rapid urbanisation following the Industrial Revolution, cycling was especially valued for the way it opened up the countryside to the huge numbers now based in towns and cities. It was, admittedly, not alone in doing this. From the 1860s onwards, cheaper train fares and improved rail services had enabled more city dwellers to take day trips to rural areas. These developments greatly encouraged the growth of rambling, and short walking holidays in the Lake and Peak Districts became increasingly popular towards the century's end.[39]

The bicycle was unique, however, in giving urban men and women a personalised mode of transport by which they could swiftly and, by the end of the nineties, cheaply experience country life. The hugely beneficial impact this had on people's day-to-day lives is revealed in any number of contemporary accounts. Describing a stop he had enjoyed during an early morning ride, one Clarion club member recounted:

> I merely sat there, drinking in the sweetness of the view, the bright morning sunlight and basking in the warmth. The memory of that scene, and of many others I have viewed in my cycling peregrinations, will long remain with me, to lighten the burden of monotony in these dreary Lancashire towns.[40]

By enabling experiences such as this, cycling tapped into the widespread romanticism with which the British countryside was regarded in this period. From industrialisation's early beginnings in the late eighteenth century, it had fostered a reaction which idealised and celebrated rural spaces. Through the writings of romantic poets such as Wordsworth and Shelley, the countryside came to be viewed as a natural counterbalance to the pollution, overcrowding and pressures of city living. The continued influence of their philosophies can be seen in pieces such as 'The Joys of Cycling by a Middle-Aged Woman', which appeared in the *Manchester Guardian* in 1895. Here the writer rejoiced:

> Out of the heart of a busy manufacturing town, with its perpetual hum of men and machinery; out of its grimness and smoke, its bustle and unrest, its struggles and failures, into pure spaces filled with peace and calm, into quiet nooks fragrant with meadowsweet and briar, woodruff and honeysuckle, into valleys whence one sees the clear, still mountain tops, and realises anew a 'central peace subsisting at the heart of ceaseless agitation'; and this to be had so easily, requiring only will, energy, and a bicycle.[41]

Also giving the bicycle significant popular appeal was a broader national obsession with British history. The Romantic movement had idealised not just nature, but also the communities and ways of life it encouraged. The Middle Ages, in particular, were romanticised as an age when society was held together not by consumerism or the pursuit of profit, but chivalry and noble deeds, where even the poorest peasants were strong, well fed and contented.[42] This view of a heroic pre-industrial England was inextricably tied into the early nineteenth-century novels of Sir Walter Scott, who enjoyed a lasting hold on the Victorian imagination. To use Mark Girouard's summary of Scott's evocative take on the past:

> He described castles complete with drawbridges, iron-studded gates, and portcullises; smoke-blackened, armour-hung halls; Christmas feasting, with Yule logs and Lords of Misrule;

maypoles and the whole concept of Merrie England; tilts, tournaments, and knights with ladies' favours pinned to their helmets, Richard Coeur de Lion, Robin Hood and his merry men, Arthur and the Knights of the Round Table.[43]

One of the key attractions of cycling among late nineteenth-century men and women journeying out on their 'steeds' was being able to visit places associated with these romanticised views of the past. As we saw with the boyish transformation of 'The Unchained' when cycling in South Devon, childhood tales of 'old England's watchdogs' performing feats of derring-do could still fire the imaginations of travelling cyclists. Although he was not quite as exuberant in his writings, the diaries of Eton headmaster and noted essayist Arthur Benson reveal how he became similarly enthralled by the historical sites he visited during cycling holidays. This included Tewkesbury Abbey:

The chapels and chantries are delicious, and there are many moving historical things. The Duke of Clarence (drowned in the butt of malmsey) lies in a vault with his duchess; Hugh Despenser, favourite of Edward II – a young man, Duke of Warwick, King of the Isle of Wight, Jersey and Guernsey who died at 21 – lies in the choir. A beautiful Early English tomb in the South East choir aisle, *most* beautiful.[44]

The fervent interest Benson took in antiquated buildings highlights another outlook common to the Victorian period. For at a time when people were urged to improve and develop themselves in their leisure hours, it was widely held that 'everyone must have their hobbies'. In Benson's case, fascinated by church design he planned routes which gave him frequent opportunities to study their architecture. During a cycle trip to Howden, a Yorkshire town containing a run-down minster, he recorded how he was still

> … simply stupified [sic] by the beauty of the church. The West front with fine open pinnacles, and the soaring tower with preternaturally long belfry windows, all of a grey creamy stone… We strolled about: the ruined choir full of graves, the chapter-house a lovely little roofless octagon with beautiful panelled work, mouldering windows, lovely diaper. The sky like sapphire blue through every opening.[45]

While not all shared in Benson's passion for old churches, cycling's 'adaptability to the pursuit of hobbies' was one of its most lauded features as a pastime. There were all manner of special interests that could be enhanced through weekend excursions that took men and women into new environments. Although archaeology, botany, geology, zoology and sketching were all established cycle hobbies, no stronger relationship existed than the one between the bicycle and the camera.[46]

Two popular new technologies both benefited from major developments at the end of the nineteenth century. In the same year that Dunlop would revolutionise cycling by inventing the pneumatic tyre, the American George Eastman perfected the handheld Kodak camera, which used roll film instead of dry plates to record images. After the foundation of the Eastman Kodak Company in 1892 his invention transformed photography, being the first affordable mass-produced camera that could be used with no formal training.[47] Easily fixed onto bicycles, these cameras meant that cycling and photography went hand in hand. A *Cycling* editorial observed:

The wheelman possesses chances of securing 'snapshots' of the most beautiful land and sea scapes, historic castles and abbeys, interesting incidents of town and country life, and the thousand and one odds and ends which the photographic artist so often yearns to capture, but has not the opportunity.[48]

As we have seen on several occasions, the camera was also used to record scenes of considerably less value to 'the photographic artist', namely members of clubs pulling poses as they were 'snapped' during mid-ride breaks. But although they were the regular subjects of photographic pieces, smiling faces and relaxed attitudes do not tell the full story of clubmen's relationship with the camera. One article in the gazette of the Hull St. Andrews Cycle Club described how, during a tea-time stop:

We should have had an instructive lecture from Donnison [a club member] on the art of 'developing', but just as he was explaining to Drewery the mysteries of Hypo-something or other, he received a shower of shrimps from all quarters, which effectually put an end to what would undoubtedly have been an interesting speech.[49]

That the unfortunate Donnison received such a barrage (it is hard to tell whether the writer was sincere in his final remark), reflects how clubs did not always welcome members with a keen photographic eye. While they were happy to stand around for the occasional group photo, regularly stopping rides so more artistically minded members could indulge their hobby met with little enthusiasm. As an article in *Cycling* forewarned:

We cannot always enter with interest into the pursuits of other men, any more than we can all expect to lie on the same bed, and non-photographic members will consider it irksome to wait while some member unstraps his 'box', and obtains a picture; and, as the selection of pictures requires an unhampered eye, and sober judgement, the operator is not assisted in his work by having a lot of fellows dogging his footsteps at every turn.[50]

It was therefore natural for dedicated cycling photographers, and others who enjoyed hobbies that required frequent breaks from riding, to make their excursions either alone or with a select group of fellow enthusiasts. Eager to find new spaces in which they could practise their special interests, they were often found journeying out on longer-distance 'tours'. Typically taking place over the course of a few days, 'touring' was a more adventurous mode of cycling celebrated for the way it enabled riders to explore new scenery and surroundings. In an article for the *Westminster Review* titled 'Bicycle Tours – And a Moral', E.H. Lacon Watson exalted:

I know of few pleasanter experiences than this, and no better method of making a large acquaintance with country charms... There are a thousand quaint, old-fashioned spots to visit, and unexpected recesses to which no railway has yet penetrated in this land of ours. Here is employment enough for the lees of a lifetime; one may go a pilgrimage still, like Tom Jones and the ancient heroes, through all the inns of the country. It is strange if you do not meet with an adventure or two, even in these prosaic days, that will be pleasant to recall hereafter over the walnuts and wine.[51]

24 Four couples photographed with their bicycles outside a monastery (date unspecified).

25 Mixed-sex group pictured with their machines during a mid-ride break (date unspecified).

As can be deduced from its title, this form of the pastime was inseparably tied into Britain's largest cycling association – the Cyclists' Touring Club. Initially established as the Bicycle Touring Club in 1878, during the closing years of the century the C.T.C. enjoyed an unprecedented surge in popularity, with 60,000 members at its peak in 1899. On the back of such widespread support it played a crucial role in promoting and protecting the interests of cyclists generally. From its earliest years it had successfully campaigned across a range of issues, from prosecuting over-aggressive policemen to fighting a bicycle tax that would have disproportionally affected poorer enthusiasts.

The C.T.C.'s main attraction to those who joined it, however, was the huge amount it did to facilitate long-distance touring. Tourists approaching the top of a steep hill in this period were likely to see a conspicuous C.T.C. sign warning unfamiliar riders that 'this hill is dangerous' and needed to be descended slowly. It also produced a series of road books which provided maps, suggested routes and insider tips that were an enormous help to visiting riders. And at the end of a long day's travelling, C.T.C.-approved inns and hotels ensured cyclists could easily find hospitable accommodation, with special rates negotiated for those affiliated with the organisation.

Compared to many local clubs, who had limited memberships to ensure they remained home to men of a certain class, the C.T.C. was a very open institution. Any cyclist who could afford its low membership fees of three shillings and sixpence annually was encouraged to join, and instantly gained access to the same privileges as wealthier members. Loudly raising its voice against proposals for a cycle tax, the C.T.C. also helped ensure that cycling remained an affordable pastime open to different classes of society, with its gazette chastising wealthier members who were 'disposed to ally themselves with the enemy' by supporting such a measure.[52]

Nevertheless, with the heavy focus it gave to cycle touring, it was to be expected that the C.T.C. acquired a reputation for being 'par excellence the club for professional men'. This was in part due to the costs which came with a cycling holiday – even with discounts on accommodation, the money needed for board, food and additional railway travel could quickly add up over a few days. In the forms promoted by *Cycling* and the C.T.C. gazette, 'touring' was also strongly associated with riders of a certain age and class.

The philosophy behind this mode of cycling was not covering as many miles in a day as possible, but engaging in hobbies and other activities that developed body and mind alike. As such, tourists were typically presented as being refined individuals like Benson, interested in the history, nature and architecture of the places they visited; in other words, the direct opposite of unruly 'scorchers'. 'If these young men enjoy cycling', commented R.J. Mecredy on those who could be seen 'swarming out of the large towns on the first and last days of the week', then 'the cultured and educated will enjoy it ten-fold', as he celebrated a cycle tour as an activity

> ... in which the cyclist is able to note all the features of a country in a way no other tourist can, and in which he is brought into contact with interesting people, curious scenes and strange customs, and comes back with a deeper knowledge of mankind, a more charitable feeling towards his humbler brethren, a mind stored with material for months of thought and reflection, and a body healthy and strong.[53]

Riders who shared in Mecredy's views often ventured overseas for their cycle tours, which allowed even greater opportunities to experience 'interesting people, curious scenes and strange customs'. At a time when cruise ships were providing frequent and ready access to 'the Continent', cycling provided an ideal means for British tourists to experience the scenery and ways of life of their European counterparts. With the C.T.C. also doing much to support overseas touring with their 'Foreign Handbook' – containing maps, information on railway and steamboat charges, useful foreign phrases and recommended hotels – overseas touring became increasingly accessible and popular during the 1890s.

In reading accounts of British cyclists overseas, however, it is soon apparent that finding material for thought and reflection was far from the sole purpose of their expeditions. Although we might celebrate the opportunities the bicycle created for travel and exploration, it is worth taking a moment to consider things from the perspective of those who lived in European countries during this period. For these developments ensured they became increasingly familiar with that all-too-recognisable modern phenomenon: the British tourist abroad.

'Can't stand those beastly French'

Of all the overseas locations frequented by British cyclists, none was more popular than northern France. Easily accessible by steamboat, the area was famous for its roads and rural scenery, while also offering the opportunity to travel down and experience the grandeur and opulence of Paris. 'I can confidently assert that no unprejudiced Englishman can return from such a tour as we had without a feeling of respect and esteem for the great French nation,' commented a tourist who had visited Normandy and Brittany, as he detailed the magnificent countryside, culture and ways of life of the French people he and his companions had met.[54]

As might be expected, however, thousands of British tourists descending on this part of the world did not always produce such a beneficial effect on Anglo-French relations. 'If ever such a calamity should occur as a war between England and France,' gloomily pronounced a *Cycling* article in 1898, "tis the English tourist who should pay the bill.'[55] Such a dark assessment was in no small part linked to the fact that those who went cycling in France were often not the cultured cyclists associated with the C.T.C. Rather, they were boisterous male riders in much the same mould we have encountered previously. Their approach to 'touring' was consequently informed far less by the aim of broadening their horizons than by rooting out rather more basic sources of pleasure and release.

We can see this in the activities enjoyed by three Stanley Cycling Club men during a tour to France in 1899. After the club gazette described how they had been 'peppering some of ours with postcards of a particularly chaste design' during their excursion, the following month one of those present sat down to write a full account of their travels. Unsurprisingly, alcohol featured heavily. With the belief that 'to my mind, the best thing to cycle on in France is white wine and water', the writer and his two companions also paused their rides to drink absinthe in local cafés, while the crew's most junior member, Jack Carrodus, was commended for his 'meritorious conduct in connection with the absorption of French beer'.[56]

When visiting towns and cities, the trio also eagerly took opportunities to experience French music halls, otherwise known as cabarets. Showcasing acrobats, singers, magicians and dancers (the modern form of the can-can was born in French cabarets such as the Moulin Rouge), the writer and his companions spent much of their time enjoying 'finely developed' female singers and stomach dancers. At one of these venues the writer detailed how a 'damsel from the stage kidded us to stand her a drink… Jack was particularly struck with her, and we had almost to use force to get him out of the place, in fact, I came to the conclusion that he is a man more amorous than the average'.[57]

Although Stanley Club members may have eagerly taken in this aspect of French culture, as with other British cyclists they were not inclined to extend this open-mindedness to other areas of national life. It was detailed with great pride how they had 'invaded' the country, with the article on their adventures titled 'The Land of the Darned Mossoo'. That travelling around France was liable to bring out the jingoistic side of visiting cyclists is demonstrated by a tourist who wrote into *Cycling* reprimanding his fellow 'Britishers' for their 'rudeness of manners and air of affected superiority' when in the country:

> It makes one quite ashamed of one's countrymen, when one sees a group of English cyclists enter a quiet country town in France, shouting 'Rule Britannia' and other patriotic songs, which we all love to sing and hear sung on every fitting occasion, but which there is no need to din into the ears of peaceful foreigners.[58]

Unfortunately, relations rarely improved when cyclists moved away from English and tried using rudimentary French to communicate with locals. Writing in 1895, one cycle journalist described how he and a friend had been 'personally conducted' on a recent tour of Belgium by their acquaintance 'Tomkins'. Described as a 'fashionably dressed young man of gentlemanly exterior', Tomkins had been selected to act as the 'guide philosopher' for their tour due to his supposedly 'intimate acquaintance with the French tongue'. This was put to the test when he was forced to speak to a railway porter after losing his rain jacket at a railway station, with their conversation being recounted as follows:

'When I was at Antwerp,' began Tomkins, 'Quand j'etais à Anvers, j'avais un mackintosh. Je n'ai pas encore un mackintosh.'

'Oui, m'sieu!'

'Où est mon mackintosh?'

'Oui, m'sieu!'

'Je dis, où est-il?'

'Oui, m'sieu!'

'What do you mean? Kesker-vous – Oh! dash it all! Look here! I say – parlez-vous Français?'

'Oui, m'sieu – oui, oui!'[59]

After carrying on for a good while longer, the writer tells us that the railway porter 'evidently began to regard Tomkins as a harmless lunatic, and seemed surprised we did not take more care of him'. Not only struggling to locate misplaced waterproofs, touring cyclists also ran into difficulty when trying to communicate that most basic of requests for the 'Britisher' abroad: ordering drinks at bars. One *Cycling* article detailed the scene that unfolded after a 'dust-begrimed Briton' swaggered into a Normandy inn and sought to order a drink:

Cyclist: 'Oh, well, er-er-mosoo, avvy voo-er-*bitter* – 'arf o' bitter. Comprenny?'

Bartender: 'Ah, vous parlez Français?'

Cyclist: 'Eh, oh, yes, er-I mean *oui* – ere that is, *après une mode*.'

At this point it was explained that 'he had a sort of remembrance that *mode* was the French for *fashion*, as this was his rendering of "after a fashion". However, the phrase book set him right'. Further to their drinking and absence of French, our picture of a type of rider who was 'painfully evident when one was abroad' is completed later in the article as the writer described his behaviour upon entering the 'gay city' of 'Parree'. He would walk around 'the aristocratic promenades' in 'a dusty sweater, with hair unkempt, a dirty pipe in his mouth and his hands in his pockets' before entering cathedrals in the same fashion and speaking loudly in English. Ending his portrayal, the author remarked how this class of tourist 'mostly gets "broke" about a week before his time is up and has to come home again and tell everybody that he couldn't stand those beastly French'.[60]

Of course, not all British cycle tourists behaved in such a manner. Many lived out the touring ideals explored previously, using their cycling excursions as opportunities to develop themselves and appreciate new cultures and ways of life. This is acknowledged by the same author, who declared that there also existed cyclists who were prepared to 'put up for the night in a village where there isn't a single music hall just because it's prettily situated' and concluded:

> They have, when they get home, seen some lovely country, laid in a grand store of health, improving alike their minds and bodies... They are actually foolish enough to find the natives courteous and kind, and waste all their time looking at the scenery, the ruins, cathedrals and other such rot, instead of having 'a good old rorty time'.[61]

On the one hand subscribing heavily to Victorian ideals of self-improvement, on the other eager to have a 'rorty time', it is fair to say that both types of cycle tourist were unique to their period in history. But the basic issue that divided them, namely whether their holidays were a means of experiencing new cultures and ways of life, or seeking simple release and gratification, remains very much with us to this day.

Indeed, the raucous and high-living cyclists explored above are most probably the ones who would be most at home in the modern world. After all, they were forced to holiday in a world where only a minority of those in foreign countries spoke English, when heavy drinking was even more strongly frowned upon than it is today, and when speaking loudly and being inappropriately dressed in a cathedral didn't make you fit in, but stand out. Individuals such as the Stanley Cycling Club members who meritoriously absorbed French beer and visited music halls might then be paid a compliment given to all great historical figures: they were, in their own peculiar fashion, well ahead of their time.

Londonderry, Bacon and 'New Women'

While it is easy to focus on the differences which divided touring cyclists, it must be noted that there was one thing that united them: they were typically portrayed as male riders. In the same way as it was all-male clubs that most clearly challenged respectable behaviour when out on their weekend runs, so it was that rowdy and obnoxious touring cyclists were viewed as an exclusively masculine phenomenon.

The picture is not quite so clear cut for touring riders at the other end of the scale. As we have already seen with the Pennells, it was common for wives to accompany their partners on cycling tours in which they travelled at a steady pace and found inspiration from the cultural sites and scenery they experienced. Compared to consuming large quantities of alcohol, visiting music halls and being unashamedly loud-mouthed in public, these touring activities were ones which could be far more easily reconciled with prevailing understandings of 'womanly' behaviour.

But it is telling that when describing tourists, writers nearly always pictured a male cyclist. Lacy Hillier's portrayal of the ideal tourist as a 'hard road rider, rather dusty, a trifle unkempt... his bronzed face is fine drawn, he looks lean and wiry' clearly had just one half of the population in mind.[62] This is symptomatic of a society in which long-distance travel and exploration, whether in overseas territories or in the pages of boys' adventure stories, was overwhelmingly associated with men. That the swashbuckling male explorer fed into images of the adventurous touring cyclist is evident in images of the English-born Thomas Stevens, who from 1884 to 1886 pedalled his machine some 13,500 miles and became the first cyclist to travel around the world.

By contrast, contemporaries would have had serious difficulties in picturing a woman in such a role. Widely held beliefs about women's weaker physical and mental make-ups, not to mention their defining roles as wives and mothers, severely undermined the notion of an autonomous female tourist. The attitudes more adventurous women riders had to contend with can be seen in an article in the *Lady Cyclists' Association News*. Here a member narrated a tour she had undertaken in Northern France, alongside a party that included several married men and a solitary unmarried one:

It was a little hard to be misunderstood by the Unmarried One. It began through my refusal to burden him with the wheeling of my bicycle up a gentle incline. 'Why not?' he asked in amazement... 'Because I never do,' was my reply... 'Of course,' he said, 'it would be an insult to suspect *you* of being tired.' 'Well,' he explained, 'frailties of that kind only belong to the feminine woman, you see.'[63]

Proving herself to be a highly capable cyclist, and later accepting a cigarette offered to her in jest, the writer of the piece happily took opportunities to challenge the attitudes of the 'Unmarried One'. By writing up accounts of their excursions, she and other female cyclists sought to spread the message that long-distance cycling was something that women could participate in just as well as men. And although British riders featured prominently in this number, perhaps the greatest example of all from this era is an American who, ten years after Stevens, achieved international fame by becoming the first woman to cycle around the world: Annie Kopchovsky.

On the surface, Kopchovsky was far from the most likely person to achieve such a feat. Born in Latvia to a Jewish family, she moved to America in 1875 and, by 1894, had married and was living in Boston as the mother of three young children. In her mid-twenties and having never ridden a bicycle, she was apparently stimulated to make her round-the-world journey by a bet made by 'two wealthy clubmen of Boston' who wagered $20,000 to $10,000 that no woman could cycle around the world in fifteen months. An additional condition was that she must start penniless and, accepting no donations or handouts, also earn $5,000 over the course of her ride.

As a married woman with children and a complete lack of cycling experience, it is unclear how Kopchovsky came to be selected as the person who would settle the dispute. Not only this, but in a city and country known for its anti-Semitism, her surname marked her out as someone likely to receive unwanted attention and hostility. Killing two birds with one stone, Kopchovsky resolved this issue by taking on the surname 'Londonderry' from the start of her journey. The title was taken from New Hampshire's Londonderry Lithia Springs Water Company, which was the first of many businesses to sponsor Kopchovsky. As well being paid $100 to take on the name of their company, it was agreed that Kopchovsky would attach a 'Londonderry' placard to her rear wheel, ensuring further publicity for the business. Achieving a considerable return from their investment, Annie carried the title with her for the rest of her journey, and it is by this name that she has since become widely known.

Setting off from Boston in June 1894, her initial plan was to head west across America to San Francisco, and from there catch a boat across the Pacific. By the time she had reached Chicago in late September, however, it was apparent that she would not reach her destination before winter snow cut off the mountainous route in between. Aware of how her progress had been hampered by the cumbersome women's bicycle she had been riding, it was at this point that Londonderry abandoned her original machine for a far lighter men's model, and traded in her long skirt for a pair of knickerbocker trousers.

She also adjusted her route, travelling back to New York City, from where she took a steamboat and, continuing an earlier theme, headed to northern France. After a three-week stopover in Paris, she proceeded to travel down the country by a combination of bicycle and train, reaching Marseille two weeks after leaving the capital. Here a huge crowd witnessed her spectacular arrival, as she entered the city with one bandaged foot propped up on the handlebar (an injury apparently sustained during a highway robbery), and the Stars and Stripes proudly fluttering from the back of her machine.

Despite the interest as she was generating, with only eight months to return to Boston, Londonderry seemingly had little hope of meeting the deadline set by the bet. Fortunately, the arrangement did not specify a minimum cycling distance, meaning the rest of her journey back to America was undertaken by steamship. After travelling across the Mediterranean, she passed through the Suez Canal before sailing to Yokohama from where she returned to the east coast of America.

Regularly stopping on route, she enjoyed cycling day trips at various ports of call, such as Alexandria, Colombo, Singapore, Hong Kong and Shanghai, where she also acquired evidence of her visit through the signature of the United States Consul. Upon reaching San Francisco in March 1895, Londonderry again used a combination of cycling and trains to head back across the country, arriving back in Chicago on 12 September 1895, just days before the fifteen-month deadline elapsed.

While some were inclined to observe that Londonderry had travelled more 'with' a bicycle than on one, her homeward arrival met with great fanfare and acclaim. Not only had she completed the journey in time, but she also possessed an incredible capacity for self-promotion that had enabled her to generate the $5,000 specified at the beginning of her journey. As well as selling advertising space on her bike, she earned significant sums by giving widely attended lectures on route. Alongside unlikely stories such as risking death while hunting tigers with German royalty in India, she altered her background story as the situation demanded: during her time in France she presented herself variously as an orphan, a wealthy heiress, a Harvard medical student and the niece of a US senator. After public talks, she also sold promotional photographs, silk handkerchiefs, souvenir pins and autographs, which were eagerly snapped up by an awestruck crowd of onlookers.

It must be admitted that Londonderry's ability to mix truth with fiction was no small part of the reason why her ride achieved such renown and recognition (it may well be that the original bet did not actually occur, but was another publicity stunt). Even without these qualities, however, she would have unquestionably remained a worldwide sensation. A confident, self-made woman who left behind her domestic duties to cycle around the world in rational dress, she offered a tremendous challenge to gender norms prevalent on both sides of the Atlantic. After finishing her ride, Londonderry and her family moved to New

York where she briefly took up a job as a writer for the *New York World*. Her first piece was an account of her round-the-world journey, which began with the line, 'I am a journalist and a "new woman", if that term means that I believe I can do anything that any man can do.'[64]

There was little competition when it came to female tourists who could match the scale of Londonderry's journey or generate interest on a similar scale. Nevertheless, there were plenty of other women riders who achieved impressive long-distance rides. A prime example of this is N.G. Bacon, an individual whom we have already encountered on several occasions. She was a keen touring rider, one of her most ambitious journeys seeing her cycle over 1,200 miles during a month-long tour of the United Kingdom. Writing up an account of her journey in the *Review of Reviews*, she detailed how her route had taken her from London to Glasgow via York and Durham, before the return leg of her journey took her back home through the Lake District, Wales and the Midlands.

Notable not only for the distance travelled, Bacon's ride also achieved interest for being 'the first extended solitary cycling tour undertaken by a young woman in knickerbockers'. As well as riding in the costume, Bacon kept it clearly on show when visiting cathedrals, recording with pleasure how she was received 'most courteously' by vergers and paid minimal attention by other visitors. Travelling seventy to eighty miles a day, she carried out a journey which contained all the elements commonly associated with touring, from overcoming the elements and uphill climbs, to arranging overnight accommodation and fixing mechanical problems. Alongside this were the pleasures of experiencing the 'majestic splendour' of the landscapes she travelled through, not to mention satisfying the hunger inevitably created by long-distance cycling. After a particularly hilly journey through the Lake District, Bacon recalled:

I was ravenous when I reached the hotel. I had spent four hours on the mountains. The tugging of my machine to get it up the hill, and the dragging of it back to prevent it being hurled too quick down the steep decline, had brought all my muscles into play, and the waitress seemed afraid of me as I looked at her so hungrily.[65]

While she did not state it so explicitly, Bacon's article carried much the same message as Londonderry's assertion that she could 'do anything that any man can do'. As we have seen before, Bacon published many other articles which strongly advocated women's involvement in cycling. What is particularly notable about these pieces, however, is that they did not present the pastime as being a straightforward source of freedom and independence. Instead Bacon gave a more nuanced picture, focussing on how cycling into the countryside would help women develop the broader attributes required to go out into the world and live autonomously. As she stated in an essay titled 'The Advantages and Pleasures of the Pastime' in the *Girls' Own Paper*:

To enjoy ourselves to the full, we need not have special literary, artistic, philanthropic or athletic powers, but we need only be the average girl, for there must be something latent in our nature which cycling will revivify… 'To paddle her own canoe', and do it wisely and well, requires the concentration of her best powers. Why should not these powers be awakened, and brought into play?

Going into greater detail, Bacon argued that overcoming difficulties while riding would allow her reader's 'masterful nature' to assert itself:

Is it not wiser to strengthen this inclination we have, undeveloped though it may be, to control, conquer and have
dominion over something, rather than allow it to run to
waste, or to be an inconvenient nuisance to ourselves and to
others? A masterful spirit may be utilised by a glorious spin
on the cycle, by which we are able to show our dominion over
hills and dales, the elements and space.[66]

Across different aspects of the pastime, from club life and day trips into
rural areas, Bacon paid close attention to cycling's potential for aiding
the growth of qualities which had traditionally been 'undeveloped' in
women. As previously discussed, these processes were not taking place
in isolation, but were closely linked to middle-class women's widening
horizons in other areas of life. From increased access to education
and entry into the workplace, to their participation in sports and
other outdoor activities, widespread acceptance of female cycling was
underwritten by these much wider social changes.

Cycling's unique features as a pastime also caused these broader developments to be reinforced and accelerated. Further to the
self-sufficiency it encouraged, cycling vividly highlighted the satisfaction
that could be found when developing qualities not traditionally seen as
'womanly'. It not only challenged arguments that women lacked the
qualities to be self-determining, but helped establish this as a virtue
worth fighting for. As Bacon asked at an earlier point in her article:

How many of us really know the powers of our nature? But
the cycle appeals to our active propensities by affording us
a means of whizzing through the air, satisfying our physical
nature by rapid locomotion; it awakens our mental nature
by compelling us ever to be on the alert for danger ahead,

and soothes, delights, and enlivens our appetite for the beauties of nature.[67]

During this period no other popular activity came close to capturing the excitement and opportunities which could be seized when dictating how you lived your life. 'Whizzing through the air' encapsulated both the freedom to act autonomously and the meaning and happiness which could subsequently be found through this. For the many women who took to cycling, it was quite natural to share in Bacon's philosophy that 'to be free is to live, in the fullest sense of the word, and until a girl realises what freedom is, it is not possible for her to be womanly'.[68]

Being so closely involved with these broader historical developments, the story of late nineteenth-century cycling is, as we have seen in each of the previous chapters, closely entwined with women's history. Since women faced considerable barriers and obstacles not experienced by their male counterparts, the opportunities created by cycling were especially significant for them. As one newspaper correspondent observed:

To men the cycle has been an unquestionable boon. But after all, men had a fair share of fresh air and country pleasures before the advent of the wheel. To women it has brought a new life... wider... freer, and more delightful than was dreamt of before its coming.[69]

The sense of liberation this invoked can be seen in the many highly personal accounts of those who experienced these developments. At the beginning of her article 'The Joys of Cycling by a Middle-Aged Woman' in the *Manchester Guardian*, the author of the piece asked:

I wonder if others have felt, as I have done since I took to cycling, that the old nature that one thought had long since been swept away or crushed out by the care, monotony and pressure of work and duty, was there all along... We have had many pleasures in the way of travelling, but we have never yet experienced such exhilarating enthusiasm or such complete recreation... The woman who is neither strong nor young can throw herself free for a time into all that invigorates and renews, and... find that contact with nature and humanity which enriches and emancipates.[70]

When it came to freedom and emancipation, the bicycle rightly enjoyed its strongest association with women. As we have seen numerous times, however, the pleasures of cycling were not limited to any one age group, class or gender. From the wealthy clubmen who took such delight in revelling 'as they did in the heyday of their youth' to the writer above who now believed that 'age is a matter of feeling and not of years, and that cares can sit lightly if the heart keeps young', liberation was fundamental to the bicycle. Indeed, perhaps the best way to make sense of the developments it brought about is to imagine experiences like the one above being repeated thousands of times over, and allowing such profound and human pleasures to speak for themselves.

Or you could go full circle and simply say this: Arthur Balfour had every reason to get carried away.

Conclusion:
Seeds of Life

If, in the words of A.J.P. Taylor, 'history is at bottom simply a form of story-telling', you could hardly ask for a better subject to bring to life than nineteenth-century cycling.[1] Not only is there a fantastically rich source of primary material, which vividly details how the bicycle enabled all kinds of dreams, passions and human capabilities to be lived to the full, but, from the birth of velocipedes in the mid-1860s to the craze of the 1890s, these accounts also fit into a wider developing narrative.

This is not only true of bicycles themselves, and the improvements which brought about pneumatic-tyred safeties. As we have seen, during this time cycling's place in the popular imagination also underwent a rapid upward trajectory. A machine which, for the first time in human history, made a practical form of personalised transportation widely available, the bicycle stood alongside other new technologies such as the telephone and electric lightbulb in its obvious potential to transform daily life. But, to return right back to where we started, there was no more powerful symbol for the possibilities of the coming century than the huge numbers who took to the pastime during the 1890s.

This was in no small part because of the immediate freedom and mobility that could be found through the bicycle. As we first saw with

Balfour, cycling was especially celebrated for helping tackle an age-old Victorian problem: more and more people living in polluted and over-crowded cities. Following Balfour's speech to the National Cyclists' Union, the *Daily Mail* dedicated a glowing editorial to the pastime:

> The greatest invention is the bicycle. It supplies a means of counteracting the worst results of a highly-developed civili-sation – the concentration of a vast population in immense cities, and the tendency to physical degeneration which always threatens the dwellers in towns. It has, too, been a civilising and humanising influence. Thousands who never saw or knew the country ten years ago now week by week revel in fresh air and learn to love nature at her best.[2]

But more than this, cycling tapped into wider cravings for emanci-pation and new physical sensations. Most especially with the 'New Woman', it symbolised a freer and more liberated society, no longer constrained by the conservative social norms that had exerted such a powerful influence during the Victorian period.[3] Appealing to multiple interpretations of the word 'progress', by the closing years of the century the bicycle had come to be a byword for excitement and optimism amid fears for what the future might bring.

In no time at all, however, this remarkable story began to stutter and stall. To those living through the nineties, much of the antic-ipation which surrounded the bicycle stemmed from a belief that they had only started to unlock its possibilities. While steam had powered Britain's early Industrial Revolution, electricity now offered a vision of the future in which power, people and ideas would be transported with far greater speed than ever before. Many predicted that new sources of power would revolutionise cycling, with one forward-looking Hull St. Andrews Cycle Club member writing at the end of the 1896 club run season:

> Looking so far ahead makes one wonder what will be the order of things next year. Shall we be pushing our machines along in the ordinary way or will electricity save us the exertion? There's no knowing what things will come to be.[4]

As it became apparent that such advances would not occur, the excitement which had surrounded the bicycle quickly evaporated. A machine which by the turn of the century had grown familiar, it no longer stood for the possibilities of technological progress. After a couple of summers spent promenading around parks, aristocratic bikists grew bored with the pastime, and the motor-car became their new plaything of choice. With the bicycle's highly fashionable status now lost, newspapers did away with dedicated cycling columns, while race tracks were forced to close as they struggled to attract the crowds. Badly mismanaged during the early 1900s, the Cyclists' Touring Club also suffered a calamitous fall in popularity, with one commentator recording how its appeal had shrunk to

> ... a certain number of steady enthusiasts, who are gradually lessened by the natural causes of death, illness and infirmity, and the more accidental ones of marriage, poverty and the rival attraction of golf.[5]

This is not to say cycling disappeared after the turn of the century. Still enjoyed by large numbers, weekend excursions into the countryside remained a widely popular leisure activity. Sitting outside a pub in Croydon one Sunday in 1904, one watchful observer counted nearly

2,000 cyclists passing over a two-hour period, compared to 125 motorcars.[6] In the coming decades the bicycle also acquired its now familiar reputation as a democratic utility vehicle, whose everyday practicality was especially valuable to members of the working classes.

However, despite its remaining an integral part of people's lives, the popular enthusiasm and anticipation which had surrounded the bicycle during the 1890s was a thing of the past. While being unaffordable for the population at large, the motorcar took the bicycle's place in symbolising future possibilities in personal transport and mobility. And as this potential was commonly realised through lower prices from the 1950s onwards, many predicted that cyclists would, at least around major urban centres, eventually disappear from Britain's roads.

Fortunately, such prophecies never came to pass. Indeed, as we carry our gaze forward to the present day, there is one final question which begs an answer. How is it that the bicycle has not only survived all the changes that have taken place since the close of the nineteenth century, but remains a prominent feature of daily life, and continues to inspire positive and optimistic visions of the future?

Certainly there have been technical innovations. Recent decades have given birth to all manner of new machines, ranging from mountain bikes, BMXs, exercise bikes and, much later than initially hoped, e-bikes. These have, each in their own way, done much to change the nature of the pastime and broadened its appeal. Furthermore, with all things vintage now firmly in fashion, cycling has regained its fashionable status as a 'hipster' mode of transport, becoming a key component of modern inner-city lifestyles.

But for all the changes and developments that have occurred in the last hundred and more years, it is the continuities that really stand out. After all, while the bicycle has since acquired several new forms, machines from the 1890s remain instantly recognisable today. Not only this, they also continue to bring people and technology together in a fundamentally elemental manner. Riding a bicycle remains defined by how it not only expands our natural capabilities, but does this through a process which is driven solely by what we put into it.

As we have seen, it was this feature of cycling that made it such a symbol for female emancipation. It not only represented middle-class women's expanding opportunities and possibilities for how they lived their lives. Going out riding was also a means of laying claim to independent qualities which had previously been 'underdeveloped' in women. As one article in the *Lady Cyclist* argued:

> The tens of thousands of wheel women of this country who have demonstrated that their sex are not an inferior portion of the human family in this wonderful form of outdoor sport, have rendered untold aid to the cause of equal suffrage, by dispelling the mistaken idea of women's dependence and helplessness.[7]

Still largely a self-powered form of transport, the bicycle's wonderful simplicity has enabled it to remain a force for positive social change. Affordable and accessible, it can be easily worked into people's daily lives, helping tackle rising levels of obesity and the strain this puts on health services. Emitting zero carbon emissions, cycling has an obvious potential to reduce the 40,000 premature deaths in the UK linked to air pollution each year, not to mention the much larger issue of global warming. The economic as well as the social benefits of a healthier, greener and more active population are also compelling. Sustrans, Britain's biggest cycling and walking charity, estimates that that there is a 67p benefit to society for every mile which someone cycles rather than drives.[8] More people realising the potential of the bicycle continues to be a catalyst for social changes equalling far more than the sum of their parts.

The most obvious parallels, however, exist in cycling's accompanying pleasures and joys. That it imbues such humanity and warmth into what can otherwise feel like a very distant period reflects how we still find much the same delights in the pastime as our Victorian predecessors. From packs of club cyclists tooling out on their steeds, to the inventiveness of manufacturers feeding a frenzied appetite for fashionable Lycra, to the resurgent interest in the athletic prowess of our leading racers, to couples finding intimacy on weekend rides, to the basic pleasure of reconnecting with nature, every aspect of cycling explored in the previous chapters remains an essential feature to this day. It is quite amazing to think how much day-to-day gratification and meaning still stems from the simple act of pedalling a bicycle.

I will admit to not being the most impartial of observers. But looking back, it is genuinely difficult to think of another invention which has done, and has the potential to do, so much to facilitate individual freedom and satisfaction, while at the same time promoting changes that benefit society. As a historian, you could ask neither for a more fascinating story to tell, nor one that feels more worthwhile in its sharing. Reaching the end of his first book, another writer concluded:

> There are ideas, and ways of thinking, with the seeds of life in them, and there are others, perhaps deep in our minds, with the seeds of a general death. Our measure of success in recognizing these kinds, and in naming them making possible their common recognition, may be literally the measure of our future.[9]

The bicycle – humble, simple, and essentially unchanging – is surely just such a 'seed of life', helping orientate us towards a better future. And you don't need to take my word for it. It was, after all, H.G. Wells who once remarked:

> Every time I see an adult on a bicycle, I no longer despair for the future of the human race.

Acknowledgements

This book could not have taken the shape it has without the input of collectors, clubs and museums, to whom I am hugely indebted. Special thanks go to the Veteran-Cycle Club and Nigel Land, who has been a fantastic help in sourcing images and putting me in touch with individual members. Among these are John Weiss, Lorne Shields, Mike Radford, Roger Bugg and Paul Adams. The club's online library has also been a rich resource of primary material, without which I would have been unable to write various sections of the book.

I am very grateful to Andrew Ritchie for both reading over the manuscript and his help with images. His book Quest for Speed was invaluable in writing the chapter on racing, and I would highly recommend it to anyone wanting a more in-depth and extensively researched account of nineteenth-century cycle racing.

Several images in the book are courtesy of Beamish Museum, and I am extremely thankful to Julian Harrop for his help in sourcing them and waiving any fee for using them. The Anfield Bicycle Club and David Birchall have also been very generous in allowing me to use photographs from the club's archive, much of which is digitised and can be explored at http://www.anfieldbc.co.uk/archive.html. Thanks also go to The Elgar Foundation, Cheadle Civic Society and Sarah Christiansen for their help with images.

One of their chief pleasures of writing the book was being able to visit various libraries and archives, with The British Library being an especially valuable source of information. It has also been a joy exploring the collections of the Bristol Central Library, Hull History Centre, Manchester Central Library and York Minster Library.

In terms of secondary material, I have invariably referred to articles by Sheila Hanlon when writing about women's cycling, and I would highly recommend her website: http://www.sheilahanlon.com/. Denis Pye's Fellowship is Life provides a fantastic summary of Clarion club cycling, and Geraldine Biddle-Perry's article on 'Fashioning Suburban Aspiration' was a big help when writing about men's cycle clothing. Andrew Ritchie's King of the Road and David Herlihy's Bicycle: A History provided excellent reference points when writing about bicycle development. My thinking about the relationships between members of all-male cycling clubs were also nurtured by the thoughtful insights of Alan McCormick.

I would not have had the time to write this book if my work had not been so accommodating in allowing me to reduce my hours, so thanks goes to Bryn Shorney as well as all other members of the NHS RightCare team for their enthusiasm and interest in the project. I am grateful to Duckworth Overlook for taking on the book, and Gesche Ipsen for her time as my editor.

I must also give special thanks two truly outstanding secondary-school history teachers in Adrian Targett and Felicity Shorrock, who imparted in both in me both an early love of the subject and a confidence that I may one day write a history book myself. I met Katie Canning around the same time that I stumbled across late-Victorian cycling as an undergraduate, and she showed remarkable patience in putting up with five and a half years of my ramblings about the topic, not to mention being a constant source of fun, love and laughter.

Finally, the two people to whom I am most gratefully indebted two are my Mum and Dad. I cannot begin to quantify what their support, encouragement and advice have done for me as an individual, or indeed the quality of this book. Whether as people, parents and editors, I am eternally grateful.

And Bruce you are forever my hero.

Image Credits

Page xiv   Image 1: Reproduced by kind permission of John Weiss.

Page xxiii   Images 2 and 3: Reproduced by kind permission of the Anfield Bicycle Club.

Page 5   Image 4: Author's own collection.

Page 5   Image 5: Reproduced by kind permission of Beamish Museum.

Page 23   Image 6: Reproduced by kind permission of Beamish Museum.

Page 35   Image 7: Reproduced by kind permission of Andrew Ritchie.

Page 58   Image 8: Reproduced by kind permission of the Cheadle Civic Society.

Page 58   Image 9: Image available in the public domain, and provided courtesy of *The Boneshaker*, a periodic journal of the Veteran-Cycle Club

Page 68   Image 10: Image available in the public domain

Page 68   Image 11: Image available in the public domain, and provided courtesy of *The Boneshaker*, a periodic journal of the Veteran-Cycle Club

Page 77   Image 12: Author's own collection.

Page 78   Image 13: Reproduced by kind permission of Lorne Shields.

Page 109   Image 14: Author's own collection.

Page 109   Image 15: Reproduced by kind permission of Nigel Land and the Veteran-Cycle Club.

Page 127   Image 16: Reproduced by kind permission of Paul Adams and the Veteran-Cycle Club.

Page 127   Image 17: Reproduced by kind permission of Beamish Museum.

Page 137   Image 18: Author's own collection.

Page 153   Image 19: Reproduced by kind permission of Lorne Shields.

Page 153   Image 20: Reproduced by kind permission of Beamish Museum.

Page 191   Image 21: Reproduced by kind permission of the Anfield Bicycle Club.

Page 191   Image 22: Reproduced by kind permission of Beamish Museum.

Page 194   Image 23: Reproduced by kind permission of The Elgar Foundation.

Page 221   Images 24 and 25: Reproduced by kind permission of Beamish Museum.

Colour Image Credits

Two hobby horse prints: Reproduced by kind permission of Lorne Shields.

Henry Sandham Print, 'Ride a Stearns' and Shirk Posters: Images provided courtesy of the Library of Congress Prints and Photographs Division

Pope Columbia Poster, Terrot Poster and Columbia Valentine: Reproduced by kind permission of Sarah Chrisman.

Cycles Gladiator Poster: Reproduction rights purchased from Alamy.com

Simpson Cycles Advert: Reproduced by kind permission of Beamish Museum.

Thomas Stevens Print: Image available in public domain

Notes

Introduction

1 Harold Begbie, *Mirrors of Downing Street: Some Political Reflections* (Mills and Boon, 1920), 76.

2 *The Standard*, issue 23320, 25 March 1899, 5.

3 'King of the Road', *Clarion*, 18 October 1897, 6.

4 See Hans-Ehrard Lessing, 'Karl Von Drais' Two-Wheeler – What We Know', International Cycle History Conference (ICHC) Proceedings 1990. Veteran-Cycle Club Online Library. Retrieved from http://veterancycleclublibrary.org.uk

5 R. Ackermann, *Repository of Arts, Literature, Commerce, Manufactures, Fashion and Politics* (London, 1 February 1819). Quoted in Andrew Ritchie, *King of the Road: An Illustrated History of Cycling* (London: Wildon House, 1975), 22.

6 David Herlihy, *Bicycle: The History* (London: Yale University Press, 2004), 35.

7 *Lancaster Gazette and General Advertiser, for Lancashire, Westmorland, &c.*, issue 938, 29 May 1819, 1.

8 *Girl of the Period Miscellany*, May 1869. Quoted in Andrew Ritchie, *King of the Road*, 58.

9 Andrew Ritchie, *King of the Road*, 74.

10 David Herlihy, *Bicycle: The History*, 141.

11 Catalogue of the 'Ariel' Bicycle, issued by Smith & Starley, Coventry, 1872. Quoted in Andrew Ritchie, *King of the Road*, 95-6.

12 Charles Spencer, 'Cycling: The Craze of the Hour', in *Cycling: The Craze of the Hour* (London: Pushkin Press, 2016; first published 1877), 8.

13 *Cycling*, issue 306, 28 November 1896, 417.

14 *Cycling*, issue 26, 18 July 1891, 422.

15   Andrew Ritchie, *King of the Road*, 86-7.

16   Brian Griffin, *Cycling in Victorian Ireland* (Ireland: Nonsuch Publishing, 2006), 55.

17   *The Cyclist*, 1879. Quoted in S.H. Moxham, *50 Years of Road Racing (1885-1935): A History of the North Road Cycling Club* (Bedford: Diemer and Reynold Ltd., 1935), 2.

18   Andrew Ritchie, *King of the Road*, 118-19.

19   Robert Machray, 'Ten Years of Cycling', *The Windsor Magazine: An Illustrated Magazine for Men and Women*, June 1897, 400.

20   F.T. Bidlake, *Cycling* (London: George Routledge and Sons Ltd., 1896), 15.

21   David Rubinstein, 'Cycling in the 1890s', *Victorian Studies* 21, Autumn, 1977, 51.

22   George Lacy Hillier, 'All Round Cycling', 1899. Quoted from *The Cyclist's Companion*, ed. George Theohari (London: Think Publishing, 2007), 138.

23   *Cycling*, issue 458, 28 October 1899, 321.

24   *Nairn's News of the Wheel*, issue 11, 16 January 1895, 111.

25   For a more detailed insight see Carlton Reid, *Roads Were Not Built For Cars: How Cyclists Were the First to Push for Good Roads and Became the Pioneers of Motoring* (London: Island Press, 2015).

26   *Weekly Irish Times*, 21 September 1895. Quoted from Brian Griffin, '"As Happy as Seven Kings": Cycling in Nineteenth-Century Ireland', *History Ireland* 22, January/February 2014, 32-5.

27   *Huddersfield Daily Chronicle*, issue 7790, 20 July 1892, 4.

28   David Rubinstein, 'Cycling in the 1890s', 59.

29   For wider discussions about the Victorian middle classes and leisure see Peter Bailey, '"A Mingled Mass of Perfectly Legitimate Pleasures": The Victorian Middle Class and the Problem of Leisure', *Victorian Studies* 21:1, Autumn, 1977, 7-28.

30   *Cyclists' Touring Club Monthly Gazette and Official Record*, vol. 13, issue 10, October 1894, 299.

31   *The Times*, issue 33620, 23 April 1892, 15.

32   *Cycling*, issue 302, 31 October 1896, 304.

33   L.C. Davidson, *Handbook for Lady Cyclists* (London: Hay Nisbet, 1896).

34   Flora Thompson, *Lark Rise to Candleford* (London: Penguin Books, 2008; first published 1939), 478.

35   See for instance Kathleen E. McCrone, *Sport and the Physical Emancipation of English Women 1870-1914* (London: Routledge, 1998).

36   *Freeman's Journal and Daily Commercial Advertiser*, 30 August 1899, 4.

37   *Girl's Own Paper*, issue 886, 19 December 1896, 182.

Chapter 1

1   Sheila Hanlon, 'The Battersea Park Cyclists' Row', 12 February 2015. Retrieved from http://www.sheilahanlon.com/?p=1750

2   Myra's Journal, issue 8, 1 August 1895, 7.

3   Jerome K. Jerome, *My Life and Times* (London: Harper and Brothers, 1926), 90.

4   *Cyclists' Touring Club Gazette*, volume 14, issue 8, August 1895, 218.

5   *Hearth and Home*, issue 253, 19 March 1896, 708.

6   *Cyclists' Touring Club Gazette*, volume 14, issue 9, September 1895, 279.

7   W.F.R., 'The Ethics of Cycling Dress', *Men's Wear*, 21 March 1908, 473. Quoted in Geraldine Biddle-Perry, 'Fashioning Suburban Aspiration: Awheel with the Catford Cycling Club, 1886–1900', *London Journal* 39:3, 2014, 196.

8   R.J. Mecredy, 'Cycling for Brain Workers', *Manchester Guardian*, 5 August 1893, 7.

9   Ibid.

10   *Cycling*, issue 169, 14 April 1894, 200.

11   R.J. Mecredy, 'Cycling for Brain Workers', 7.

12   Henry Sturmey, *Indispensable Cyclists' Handbook*, 1879.

13   C.P. Sisley, 'On Clubs, Cycling Institutions, and Literature Devoted to the Sport', in *The Complete Cyclist*, ed. B. Fletcher Robinson (London: A.D. Innes and Co., 1897), 309-10.

14   Geraldine Biddle-Perry, 'Fashioning Suburban Aspiration', 194.

15   Ibid., 195.

16   Zoe Lawson, 'Wheels Within Wheels – the Lancashire Cycling Clubs of the 1880s and '90s', in *Lancashire Local Studies in Honour of Dianna Winterbotham*, ed. Alan G. Crosby (Preston: Carnegie Publishing Ltd., 1993), 126.

17   Frank Marriott, *The Black Anfielders: Being the Story of the Anfield Bicycle Club* (Prescot: T. Stephenson & Sons Ltd., 1979; first published 1956), 2.

18   Regent Street Polytechnic, *Home Tidings* (Polytechnic Gazette), 4 April 1885, 65. Quoted in Geraldine-Biddle Perry, 'Fashioning Suburban Aspiration', 196.

19   Flora Thompson, *Lark Rise to Candleford*, 476.

20   Ibid., 476-7.

21   Quoted in *Cycling*, issue 429, 8 April 1899, 260.

22   *The Princess*, 27 June 1896, 6.

23   Andrew Ritchie, *Quest for Speed: A History of Early Bicycle Racing* (California: Andrew Ritchie, 2011), 378.

24   David Rubinstein, 'Cycling in the 1890s', 53

25   *Cyclists' Touring Club Gazette*, volume 13, issue 11, November 1895, 326.

26   Tom Groom, *National Clarion Cycling Club 1894-1944: The Fifty Year Story of the Club* (Halifax, 1944), 1. 'Records of the Clarion Cycling Club', Manchester Libraries, Information and Archives GB127.O16, Box 10.

27   *The Hub*, 9 April 1898, 383. Quoted in David Rubinstein, 'Cycling in the 1890s', 49.

28   *Cycling*, issue 298, 3 October 1896, 223.

29   *New York Recorder*, 6 February 1895. Quoted in Glen Norcliffe, *Critical Geographies of Cycling: History, Political Economy and Culture* (Oxon: Routledge, 2015).

30   Nadine Besse and André Vant, 'A New View of Late 19th-Century Cycle Publicity Posters', *International Cycle History Conference*, 1994, 117-22. Veteran-Cycle Club Online Library. Retrieved from http://veterancycleclublibrary.org.uk

31   Ross D. Petty, 'Peddling the Bicycle in the 1890s: Mass Marketing Shifts into High Gear', *Journal of Macromarketing* 15 (1995), 32-46.

32   Jerome K. Jerome, *Three Men on the Bummel* (published as *Three Men in a Boat and Three Men on the Bummel*) (London: Penguin Books, 1999; first published 1900), 292.

33   *Freeman's Journal and Daily Commercial Advertiser*, 30 August 1899, 4.

34   *Myra's Journal*, issue 4, 1 April 1898, 13.

35   Jerome K. Jerome, 'Women and Wheels', in *Cycling: The Craze of the Hour* (London: Pushkin Press, 2016; first published 1897), 89.

36   George Lacy Hillier, *The Potterers Club* (London: Gale and Polden Ltd, 1900), 8.

37   *Cycling*, issue 378, 16 April 1898, 319.

38   Ibid.

39   F.T. Bidlake, *Cycling*, 96.

40   George Lacy Hillier, *The Potterers Club*, 9.

41   George Lacy Hillier and William Coutts Keppel, Earl of Albemarle, *Cycling*, The Badminton Library of Sports and Pastimes (London: Longman, Green and Co, 1887), 251.

42   W. MacQueen Pope, *Twenty Shillings in the Pound* (London: Hutchinson, 1949), 100. Quoted in Christopher Breward, *The Hidden Consumer: Masculinities, Fashion and City Life 1860-1914* (Manchester: Manchester University Press, 1999), 49.

43   Christopher Breward, *The Hidden Consumer*, 201.

44   Prospectus of the Cyclists' Touring Club, 1897. Hull History Centre, U DX113/11.

45   Price list of cycling garments from Holding & Sons, Tailors, Oxford Circus and Cheapside. Hull History Centre, U DX113/6.

46   *Cycling*, issue 225, 11 May 1895, 276.

47   Gilbert Floyd, 'The Humours of Cycling', in *The Complete Cyclist*, 175.

48   *Cycling*, issue 182, 14 July 1894, 408.

49   H.G. Wells, *Kipps: The Story of a Simple Soul* (London: Everyman, 1993; first published 1905), 104.

50   Rose Macaulay, *Told by an Idiot*. Quoted in Jeanne Mackenzie, *Cycling* (Oxford: Oxford University Press, 1981), 18.

51   Ellen Gruber Garvey, 'Reframing the Bicycle: Advertising-Supported Magazines and Scorching Women', *American Quarterly* 47, March 1995, 69.

52   Helena Swanwick, *I Have Been Young* (London: Gollancz, 1935), 164. Quoted in David Rubinstein, 'Cycling in the 1890s', 63.

53   *Cyclists' Touring Club Gazette*. Quoted in *Sporting Times*, issue 1860, 13 May 1899, 1.

54   Ellen Gruber Garvey, 'Reframing the Bicycle', 69.

55   *The Lily: A Ladies' Journal, Devoted to Temperance and Literature*, 1849.

56   'Dangers of Women's Dress', *The Rational Dress Society's Gazette* 1 (April 1888), 3.

57   *Cyclists' Touring Club Gazette*, volume 14, issue 1, January 1895, 18-19.

58   Claire Simpson, 'Respectable Identities: New Zealand Nineteenth-Century "New Women" – on Bicycles!', *The International Journal of the History of Sport* 18:2 (2001), 54.

59   H.G. Wells, *The Wheels of Chance* (1896). Retrieved from http://www.gutenberg.org/files/1264/1264-h/1264-h.htm

60   *North-Eastern Daily Gazette*, 9 April 1895.

61   *The Buckman Papers*, letter from Kitty Jane Buckman, 23 August 1897. Quoted in David Rubinstein, 'Cycling in the 1890s', 64.

62   *Jackson's Oxford Journal*, issue 7646, 30 September 1899.

63   Records of the Rational Dress League, Hull History Centre, U DX113/14.

64   *Leeds Mercury*, issue 19034, 6 April 1899.

65   *Cycling*, issue 283, 20 June 1896, 405.

66   *Berkshire Chronicle*, 30 September 1899. Hull History Centre, U DX113/5.

67   *Cheltenham Examiner*, 8 September 1897. Hull History Centre, U DX113/5.

68   Newspaper not specified. Press cuttings, 1894-1899. Hull History Centre, U DX113/5.

69   'A Gathering of the Supporters of the Rational Costume'. Hull History Centre, U DX113/10.

70   *The Queen: The Lady's Newspaper*, September 1897, 476. Hull History Centre, U DX133/5.

71   David Rubinstein, 'Cycling in the 1890s', 65-6.

72   *Daily Mail*, 17 May 1898. Hull History Centre, U DX113/5.

73   Mihaela Peteu and Sally Helvenston, 'Invention, the Angel of the Nineteenth Century: Patents for Women's Cycling Attire in the 1890s', *Dress* 32 (2005), 38-9.

74   Gwen Raverat, *Period Piece* (1952). Quoted in Jeanne Mackenzie, *Cycling*, 18.

75   David Rubinstein, *Before the Suffragettes: Women's Emancipation in the 1890s* (Sussex: The Harvester Press, 1986), 219.

76   Mihaela Peteu and Sally Helvenston, 'Invention, the Angel of the Nineteenth Century', 38-9; Mihaela Peteu and Sally Helvenston, 'Improving the Functionality of Women's Skirts, 1846-1920', *Clothing and Textiles Research Journal* 27, 2009, 45-61.

77   David Rubinstein, 'Cycling in the 1890s', 57-8.

78   *Cycling*, issue 236, 27 July 1895, 31.

79   *Cycling*, issue 111, 4 March 1893, 134.

Chapter 2

1   A.J. Wilson, *The Pleasures, Objectives and Advantages of Cycling: With Numerous Illustrations* (London: Ilfe and Son, 1887), 65-6.

2   James Hawkers, *A Victorian Poacher*. Quoted in Jeanne Mackenzie, *Cycling*, 12.

3   *Funny Folks*, issue 965, 20 May 1893, 354.

4   *Leeds Evening Express*, 22 August 1896, 2.

5   *The Times*, issue 33614, 16 April 1892, 7.

6   *The Times*, issue 33620, 23 April 1892, 15.

7   *Tottenham Cyclist. The Official Organ of the Tottenham Cycling Club*, volume 1, issue 4, June 1897, 2.

8   *Cycling*, issue 67, 30 April 1892, 232.

9   George Lacy Hillier, *The Potterers Club*, 9.

10   *Cyclists' Touring Club Gazette*, volume 13, issue 12, December 1894, 366.

11   *Cycling*, issue 236, 27 July 1895, 24.

12   *Punch*, 7 May 1892, 217.

13   *Cycling*, issue 177, 9 June 1894, 334.

14   *Cyclists' Touring Club Gazette*, volume 13, issue 11, November 1894, 315.

15   *Cycling*, issue 397, 27 August 1898, 132; *Cycling*, issue 323, 27 March 1897, 230.

16   Ken Dobb, 'An Alternative Form of Long Distance Cycling: The British Roads Records Association'. http://www.randonneurs.bc.ca/history/an-alternative-form-of-long-distance-cycling_part-3.html

17   Andrew Ritchie, *Quest for Speed*, 263.

18   Bath Road Club Album, Veteran-Cycle Club Online Library. Retrieved from http://veterancycleclublibrary.org.uk

19   Anfield Bicycle Club, Official Route Card of the 12 Hours Scratch Road Race, to be held on Monday, 3rd August, 1891. Retrieved from http://www.anfieldbc.co.uk/archive.html

20   Ken Dobb, 'An Alternative Form of Long Distance Cycling'. Retrieved from http://www.randonneurs.bc.ca/history/an-alternative-form-of-long-distance-cycling_part-3.html and *Bath Road Club Album*.

21   Anfield Bicycle Club, Reports and Accounts for the Year Ending 31st December 1897, 8. Retrieved from http://www.anfieldbc.co.uk/archive.html

22   Anfield Bicycle Club, Reports and Accounts for the Year Ending 31st December 1895, 8. Retrieved from http://www.anfieldbc.co.uk/archive.html

23   Ken Dobb, 'An Alternative Form of Long Distance Cycling'.

24   Andrew Ritchie, *Quest for Speed*, 242-53, 330-6.

25   Gerry Moore, *The Little Black Bottle: Choppy Warburton, the Question of Doping, and the Deaths of His Bicycle Racers* (San Francisco: Cycle Publishing, 2011), 33-43.

26   *Cycling*, issue 229, 8 June 1895, 336.

27   Peter Bailey, 'Ally Sloper's Half Holiday: Comic Art in the 1880s', *History Workshop Journal* 16, October 1983, 4-7.

28   *Sheffield & Rotherham Independent*, issue 12917, 17 February 1896, 4.

Notes

29   'The Cycle Champions', *The Saturday Review*, 7 August 1897, 136.

30   *Derby Mercury*, issue 9527, 11 August 1897, 8.

31   *The Hub*, 10 October 1896, 363. Quoted in Andrew Ritchie, *Quest for Speed*, 338.

32   Andrew Ritchie, *Quest for Speed*, 343.

33   For more on this see Chapter 8: 'Bicycle Racing and Modernity' in Andrew Ritchie, *Quest for Speed*.

34   Sheila Hanlon, 'Tessie Reynolds: The Stormy Petrel in the Struggle for Women's Equality in Cycle Racing and Dress', 17 April 2017. Retrieved from http://www.sheilahanlon.com/?p=1830

35   *Cycling*, issue 139, 16 September 1893, 136.

36   Ibid.

37   *Hearth and Home*, issue 210, 23 May 1895, 59; *Hearth and Home*, issue 235, 14 November 1895, 22.

38   *Lady Cyclists' Association News*, issue 23, November 1898, 23. Hull History Centre, U DX113/13.

39   *Cyclists' Touring Club Gazette*, volume 15, issue 3, March 1896, 96.

40   *Lloyd's Weekly Newspaper*, issue 2766, 24 November 1895, 11.

41   'Three Years as a Lady Racer: Some Experiences of Miss Blackburn', *The Hub*, 12 March 1898, 231. Hull History Centre, U DX113/5.

42   Claire Simpson, 'Capitalising on Curiosity: Women's Professional Cycle Racing in the Late Nineteenth Century', in *Cycling and Society*, ed. Dave Horton, Paul Rosen and Peter Cox (Aldershot: Ashgate, 2007), 47.

43   *Cycling*, issue 143, 14 October 1893, 197; *Cycling*, issue 141, 30 September 1893, 164.

44   Claire Simpson, 'Capitalising on Curiosity', 47-66.

45   'How Ladies' Cycle Races are Managed', *The Hub*, 1896. Quoted in Claire Simpson, 'Capitalising on Curiosity', 52.

46   'Interesting Bits of Information', *The Hub*, 1897. Quoted in Sheila Hanlon, 'Ladies' Cycle Races at The Royal Aquarium: A Late Victorian Sporting Spectacle', 26 January 2015. Retrieved from http://www.sheilahanlon.com/?p=1556

47   *Cycling*, issue 149, 25 November 1893, 306.

48   *Birmingham Daily Post*, issue 11681, 25 November 1895.

49   For an excellent discussion of professional female racing in this period, see Sheila Hanlon, 'Ladies' Cycle Races at The Royal Aquarium: A Late Victorian Sporting Spectacle' http://www.sheilahanlon.com/?p=1556

50   *The Hub*, 12 March 1898, 231. Hull History Centre, U DX113/5.

51   Dick Swann, 'Early Women Racers', *The Boneshaker: The Magazine of the Veteran-Cycle Club*, volume 13, issue 128, Spring 1992, 7.

52   Ibid., 10.

53   'The Latest Cycling Sensation', *Kalgoorlie Western Argus*, 8 March 1904, 41. Quoted in Philip McCouat, 'Toulouse-Lautrec, the Bicycle and the Women's Movement', *Journal of Art in Society*. Retrieved from http://www.artinsociety. com/toulouse-lautrec-the-bicycle-and-the-womens-movement.html

54   'Air-Girl's Challenge', *Northern Star* (Lismore), 7 December 1911, 4. Quoted in Philip McCouat, 'Toulouse-Lautrec, the Bicycle and the Women's Movement'.

55   Claire Simpson, 'Capitalising on Curiosity', 63.

56   'The Cycle Racing of 1893', *Wheelmen's Gazette*, November 1893. Quoted in Andrew Ritchie, *Flying Yankee: The International Cycling Career of Arthur Augustus Zimmerman* (Cheltenham: John Pinkerton Memorial Publishing Fund, 2009), 117.

57   *Indianapolis Sentinel*, 25 August 1893. Quoted in Andrew Ritchie, *Quest for Speed*, 310.

58   *Cycling*, issue 181, 7 July 1894, 396.

59   Gerry Moore, *The Little Black Bottle*, 101.

60   Ibid., 96-7.

61   'A Good State for Cyclists', *New York Times*, 30 June 1893.

62   *Cycling* Race Meeting, Crystal Palace, 20 May 1905, 23.

63   Quoted in Andrew Ritchie, *Flying Yankee*, 110.

64   *The Referee* (Sydney), 20 November 1895. Quoted in Andrew Ritchie, *Flying Yankee*, 112.

65   Andrew Ritchie, *Flying Yankee*, 112-20.

66   Steven Thompson, 'Michael, James (1875–1904)', *Dictionary of National Biography* (Oxford: Oxford University Press), May 2011. Retrieved from http://www.oxforddnb.com/view/article/101137

67   Richard Holt, *Sport and the British: A Modern History* (Oxford: Clarendon Press, 1989), 74-134.

68   R.J. Mecredy, 'Cycling', *Fortnightly Review* 50, July 1891, 88.

69   Richard Holt, *Sport and the British*, 99.

70   Robert McCrum, 'Penalty Shoot-Outs? Blame My Great-Grandfather', 3 July 2004. Retrieved from https://www.theguardian.com/football/2004/jul/04/euro2004.sport9

71   Andrew Ritchie, *Quest for Speed*, 89-97. See also Wray Vamplew, *Pay Up and Play the Game: Professional Sport in Britain, 1875-1914* (Cambridge: Cambridge University Press, 1988), 189.

72   Andrew Ritchie, *Quest for Speed*, 62-97.

73   Ibid., 92.

74   Ibid., 86.

75   *Cycling*, issue 20, 6 June 1891, 320.

76   James McCullagh, *American Bicycle Racing* (Rochdale Press, 1976), 13-16.

77   Andrew Ritchie, *A Quest for Speed*, 309-12.

78   Ibid., 354-8.

79   'Irish Cycling News', *American Wheelman*, October 1887. Quoted in Andrew Ritchie, *Quest for Speed*, 258-9.

80   *Cycling*, issue 191, 15 September 1894, 144.

81   *Cycling*, issue 153, 23 December 1893, 425.

82   Ibid., 425.

83   *Cycling and Moting*, issue 518, 22 December 1900, 512.

84   *Cycling and Moting*, issue 475, 24 February 1900, 129.

85   Andrew Ritchie, *Quest for Speed*, 258-67.

86   'Causerie de Jour – George Lacy Hillier', *La Bicyclette*, 16 October 1892. Quoted in Andrew Ritchie, *Quest for Speed*, 259.

87   Ibid.

Chapter 3

1   *Bristol Bicycle and Tricycle Club Monthly Gazette*, volume 1, issue 8, August 1897, 67.

2   *Bristol Bicycle and Tricycle Club Monthly Gazette*, volume 1, issue 7, July 1897, 60.

3   *Stanley Gazette. The Official Organ of the Stanley Cycling Club*, volume 8, issue 92, September 1899, 69.

4   *Bristol Bicycle and Tricycle Club Monthly Gaz*ette, volume 1, issue 9, September 1897, 81.

5   Walter Blake, 'History of the Pickwick Bicycle Club'. Retrieved from http://www.pickwickbc.org.uk/History-of-PBC.html

6   Canterbury Cycle Club, 'Rules and Byelaws 1893-94' (Canterbury: Cross & Jackman, 1893), 7. Although some clubs abandoned such rules during the 1890s, they continued to be enforced (minus the bugler) by many others.

7   Walter Blake, 'History of the Pickwick Bicycle Club'.

8   *Boys Own Paper*, 13 August 1881. Quoted in 'Hampton Court Bicycle Ride'. Retrieved from http://www.pickwickbc.org.uk/Hampton-Court.html

9   *London and Provincial Illustrated Newspaper*, 20 May 1876, 362. The Veteran-Cycle Club Online Library. Retrieved from http://veterancycleclublibrary.org.uk/ and *The Standard*, issue 16793, 20 May 1878, 3.

10   Walter Blake, 'History of the Pickwick Bicycle Club'.

11   Ibid.

12   *Cycling*, issue 152, 16 December 1893, 402.

13   Ibid.

14   *Argus Bicycle Club Gazette*, volume 2, issue 14, 14 November 1891, 161.

15   Paul Creston, 'Cycling and Cycles', *Fortnightly Review*, May 1894, 679.

16 *Stanley Gazette*, volume 2, issue 20, September 1893, 81. The piece actually described a visit to the camp's southern equivalent, which was held near Dorking. Copying the same formula as the Harrogate Camp, the Dorking Camp ran as a parallel event, more accessible to members from southern clubs.

17 Henry Sturmey, 'The Birth of the Harrogate Camp', in *Fun and Frolic in Cycling Camps at Harrogate &c.*, ed. J.B. Radcliffe (Leeds: Chorley and Pickersgill, 1898). Veteran-Cycle Club Online Library. Retrieved from http://veterancycleclublibrary.org.uk/. For other accounts of the evening in same publication see George Lacy Hillier, 'Cyclists Under Canvas' and Earnest Hickson, 'Past and Present'.

18 *Ariel Gazette*, volume 1, issue 3, August 1893, 25. Veteran-Cycle Club Online Library. Retrieved from http://veterancycleclublibrary.org.uk/

19 *Cycling*, issue 118, 22 April 1893, 243.

20 *Bristol Bicycle and Tricycle Club Gazette*, volume 1, issue 4, April 1897, 26 and 29.

21 *Stanley Gazette*, volume 6, issue 66, July 1897, 44.

22 *Stanley Gazette*, volume 8, issue 91, August 1899, 57.

23 *Argus Bicycle Club Gazette*, volume 2, issue 3, 13 June 1891, 34.

24 *Stanley Gazette*, volume 8, issue 88, May 1899, 29.

25 *Cyclists' Touring Club Gazette*, volume 15, issue 2, February 1896, 41.

26 *Cycling*, issue 120, 6 May 1893, 276.

27 *Bristol Bicycle and Tricycle Club Gazette*, volume 1, issue 5, May 1897, 39-40.

28 *Stanley Gazette*, volume 8, issue 88, May 1899, 29.

29 J.B. Radcliffe, 'Introduction by the Editor', in *Fun and Frolic in Cycling Camps at Harrogate &c.*

30 *Argus Bicycle Club Gazette*, volume 2, issue 14, 14 November 1891, 161.

31 *Bristol Bicycle and Tricycle Club Gazette*, volume 1, issue 1, January 1897, 1.

32 Henry Sturmey, *The Cyclists' Year Book for 1899*. Quoted in Andrew Ritchie, *Quest for Speed*, 375, and *The Bristol Mercury and Daily Post*, 13 January 1890, issue 13001.

33 *Leeds Evening Express*, 16 April 1898, 2.

34 Andrew Ritchie, *Quest for Speed*, 370.

35   *Penny Illustrated Paper and Illustrated Times*, issue 1785, 10 August 1895, 96.

36   Andrew Ritchie, *Quest for Speed*, 370-1.

37   Ancoats Wesleyan Cycling Club, 'Guidebook for the 1899 Season'. Manchester Libraries, Information and Archives.

38   *Stanley Gazette*, volume 6, issue 66, July 1897, 1.

39   *Cycling*, issue 160, 10 February 1894, 56.

40   *Argus Bicycle Club Gazette*, volume 1, issue 12, 18 October 1890, 8.

41   *Stanley Gazette*, volume 6, issue 66, July 1897, 1.

42   *Leeds Evening Express*, 12 March 1898, 2.

43   Robert J. Lake, 'Gender and Etiquette in British Lawn Tennis 1870–1939: A Case Study of "Mixed Doubles"', *The International Journal of the History of Sport* 29:5, 2011, 691-710.

44   David Rubinstein, 'Cycling in the 1890s', 68.

45   Sheila Hanlon, 'Ladies Cycling Clubs: The Origins and Politics of Victorian Women's Bicycling Associations', 15 September 2015. Retrieved from http:// www.sheilahanlon.com/?p=1889

46   *Cyclorn: Official Organ of the Hull St. Andrews C.C.*, volume 3, issue 2, May 1896, 6.

47   Sheila Hanlon, 'Ladies Cycling Clubs'.

48   *Hearth and Home*, issue 251, 5 March 1896, 646.

49   *Lady's Own Magazine*, June 1897, 23. Hull History Centre, U DX113/15.

50   'Records of the Western Rational Dress Club', Hull History Centre, U DX113/8.

51   Sheila Hanlon, 'Ladies' Cycle Races at The Royal Aquarium'.

52   *Girl's Own Paper*, issue 888, 2 January 1897, 219.

53   *Cycling*, issue 198, 3 November 1894, 258.

54   *North-Eastern Daily Gazette*, 19 October 1898.

55   'Manchester Ladies Oxford Cycling Club Rules and Programme', 1898. Manchester Libraries, Information and Archives, GB127.M522/5/1.

56   *Girl's Own Paper*, issue 888, 2 January 1897, 219.

57   Sheila Hanlon, 'Flora Drummond: The Suffragette General'. Retrieved from
      http://www.sheilahanlon.com/?page_id=1371

58   Dennis Pye, *Fellowship is Life: The Story of the Clarion Cycling Club* (Bolton: Clarion
      Publishing, 1995), 4-8.

59   Ibid., 8.

60   Tom Groom, *National Clarion Cycling Club 1894-1944.*

61   Dennis Pye, *Fellowship is Life*, 27-33.

62   Tom Groom, 'The Fifty Year History of the Club, 1894-1944', 1.

63   *The Scout – A Journal for Socialist Workers*, volume 1, issue 3, June 1895. Quoted
      in David Prynn, 'The Clarion Clubs, Rambling and the Holiday Associations
      in Britain since the 1890s', *Journal of Contemporary History* 11, July 1976, 66.

64   Sylvia Pankhurst, 'Clarion Cycling Days 1896-1898'. Retrieved from http://
      country-standard.blogspot.co.uk/2011/05/sylvia-pankhurt-clarion-cycling-
      days.html

65   Quoted in Alastair Bonnett, *Left in the Past: Radicalism and the Politics of Nostalgia*
      (London: Continuum, 2010), 76.

66   Dennis Pye, *Fellowship is Life*, 16-17.

67   Tom Groom, 'The Fifty Year History of the Club, 1894-1944'.

68   Sylvia Pankhurst, 'Clarion Cycling Days 1896-1898'.

69   *Stanley Gazette*, volume 9, issue 104, September 1900, 59.

Chapter 4

1   Quoted in Richard Anthony Baker, *British Music Hall: An Illustrated History*
      (Barnsley: Pen and Sword History, 2010), 133-5.

2   Gerry Moore, *The Little Black Bottle*, 36-7.

3   *Weekly Standard and Express*, issue 3094, 8 June 1895, 8.

4   *The Tricyclist*, 30 June 1882. Quoted in Andrew Ritchie, *King of the Road*,
      117-18.

5   A.J. Wilson, *The Pleasures, Objects, and Advantages of Cycling*, 75-6.

6   *Stanley Gazette*, volume 2, issue 22, November 1893, 95.

7   Dave Buchanan, 'The Pennells: Cycle Touring's First Couple', *Adventure Cyclist*, June 2016, 12-16 and 46. Retrieved from https://www.adventurecycling.org/default/assets/resources/20160601_Pennells_Buchanan.pdf

8   Elizabeth and Joseph Pennell, *Our Sentimental Journey through France and Europe* (London: Longmans, Green & Co., 1888), 182-3.

9   Elizabeth and Joseph Pennell, *A Canterbury Pilgrimage* (London: Seeley and Company, 1885), 50.

10   Elizabeth and Joseph Pennell, *Our Sentimental Journey through France and Europe*, 117-18.

11   P.J. Perry, 'Working-Class Isolation and Mobility in Rural Dorset, 1837-1936: A Study of Marriage Distances', *Transactions of the Institute of British Geographers*, no. 46, March 1969, 133-4.

12   Steve Jones, *The Language of the Genes* (London: HarperCollins, 1994), 315.

13   *Nottinghamshire Guardian*, issue 2487, 14 January 1893, 6; *Huddersfield Chronicle and West Yorkshire Advertiser*, issue 10350, 29 September 1900, 16.

14   *Hampshire Telegraph and Sussex Chronicle*, issue 5708, 17 January 1891.

15   *Freeman's Journal and Daily Commercial Advertiser*, 30 August 1899, 4; Florence Harcourt Williamson, 'The Cycle in Society', in *The Complete Cyclist*, 61.

16   Frances Willard, *A Wheel Within a Wheel: How I Learned to Ride the Bicycle with Some Reflections by the Way* (New York: Fleming H. Revell Co., 1895), 61-2.

17   *Freeman's Journal and Daily Commercial Advertiser*, 30 August 1899, 4.

18   David Rubinstein, 'Cycling in the 1890s', 61-2; *Sheffield & Rotherham Independent*, issue 12989, 11 May 1896, 4.

19   *Cycling*, issue 315, 30 January 1897, 59; *North-Eastern Daily Gazette*, 9 May 1896.

20   David Rubinstein, 'Cycling in the 1890s', 62.

21   *Tottenham Cyclist*, volume 1, issue 5, July 1897, 8.

22   H.G. Wells, 'A Perfect Gentleman on Wheels', in *The Humours of Cycling* (London: James Bowden, 1897), 8.

23   *Stanley Gazette*, volume 6, issue 67, August 1898, 52.

Notes

24   *Aberdeen Weekly Journal*, issue 12847, 1 April 1896, 4.

25   *Leeds Evening Express*, 14 May 1898, 2 and *The Hub*, quoted in the *Huddersfield Daily Chronicle*, 26 October 1896, 4.

26   *Freeman's Journal and Daily Commercial Advertiser*, 30 August 1899, 4.

27   *Cycling*, issue 156, 13 January 1894, 479.

28   *Bristol Bicycle and Tricycle Club Gazette*, volume 1, issue 9, September 1897, 76.

29   *Stanley Gazette*, volume 4, issue 54, July 1896, 49; *The Stanley Gazette*, volume 7, issue 80, September 1898, 61.

30   *Cyclorn*, volume 3, issue 1, April 1896, 3.

31   Ibid.

32   *Stanley Gazette*, volume 7, issue 73, February 1898, 28-9.

33   *Bristol Bicycle and Tricycle Club Gazette*, volume 1, issue 8, August 1897, 69-70.

34   *Bristol Bicycle and Tricycle Club Gazette*, volume 1, issue 9, September 1897, 81.

35   *Argus Bicycle Club Gazette*, volume 2, issue 26, 7 May 1892, 308-9.

36   Jan Marsh, 'Sex and Sexuality in the 19th Century', *The Victoria and Albert Museum*, http://www.vam.ac.uk/content/articles/s/sex-and-sexuality-19th-century/gm

37   W.J. Dawson, *The Threshold of Manhood: A Young Man's Words to Young Men* (London: Hodder and Stoughton, 1889), 89.

38   Jeffrey Weeks, '"Sins and Diseases": Some Notes on Homosexuality in the Nineteenth Century', *History Workshop* 1, Spring 1976, 216.

39   Jan Marsh, 'Sex and Sexuality in the 19th Century'; Michael Foldy, *The Trials of Oscar Wilde: Deviance, Morality and Late-Victorian Society* (London: Yale University Press, 1997).

40   *Stanley Gazette*, volume 5, issue 60, January 1897, 105.

41   *Stanley Gazette*, volume 8, issue 90, July 1899, 53.

42   *Stanley Gazette*, volume 7, issue 75, February 1898, 46.

43   John Tosh, 'Masculinities in an Industrializing Society: Britain, 1800–1914', *Journal of British Studies* 44, April 2005, 338.

44   John Tosh, *A Man's Place: Masculinity and the Middle-Class Home in Victorian England* (London: Yale University Press).

45 *British Mothers' Magazine*, 1 May 1849, 116.

46 *Argus Bicycle Club Gazette*, volume 2, issue 16, 12 December 1891, 91-2. For wider discussions see Valerie Sanders, *The Tragi-Comedy of Victorian Fatherhood* (Cambridge: Cambridge University Press, 2009) and Trev Broughton and Helen Rogers, 'Introduction', in *Gender and Fatherhood in the Nineteenth Century*, ed. Trev Broughton and Helen Rogers (Basingstoke: Palgrave Macmillan, 2007), 1-28.

47 *Stanley Gazette*, volume 1, issue 7, February 1892, 5.

48 *Cyclists' Touring Club Gazette*, volume 15, issue 2, February 1896, 41.

49 John Tosh, *A Man's Place*, 184-90; Simon Gunn, *The Public Culture of the Victorian Middle Class: Ritual and Authority and the English Industrial City 1840-1914* (Manchester: Manchester University Press), 186-98.

50 H.G. Wells, *The Wheels of Chance*. Retrieved from http://www.gutenberg.org/files/1264/1264-h/1264-h.htm

51 *Stanley Gazette*, volume 1, issue 7, February 1892, 5.

Chapter 5

1 *Cycling*, issue 105, 12 January 1893, 23.

2 Martin Buzacott, 'Elgar the Cyclist', 20 September 2014. Retrieved from http://www.abc.net.au/classic/content/2014/09/20/4088830.htm

3 J.N. Moore, *Edward Elgar: A Creative Life* (Oxford: Oxford University Press, 1987).

4 Elizabeth and Joseph Pennell, *Our Sentimental Journey through France and Italy*, 231-2.

5 H.G. Wells, *The Wheels of Chance*.

6 Michael Holyroyd, *Bernard Shaw: The New Biography* (Head of Zeus, 2015).

7 Ibid.

8 Shaw to Janet Achurch, 13 April 1895. From *Collected Letters of G.B. Shaw*, ed. Dan Lawrence (London: Max Reinhardt, 1965). Quoted in Jeanne Mackenzie, *Cycling*, 51.

9 Shaw to R. Golding Bright, 22 September 1896. Quoted in Jeanne Mackenzie, *Cycling*, 53-4.

Notes

10 'George Bernard Shaw and Bertrand Russell Crash Into One Another On Their Bicycles'. The British Newspaper Archive, https://blog. britishnewspaperarchive.co.uk/2013/05/03/george-bernard-shaw-and-bertrand-russell-crash-into-one-another-on-their-bicycles/

11 Shaw to Janet Achurch, 16 September 1895. Quoted in Jeanne Mackenzie, *Cycling*, 52-3.

12 Bernard Russell, *Portraits from Memory*, 1956. Quoted in Jeanne Mackenzie, *Cycling*, 53.

13 *Yorkshire Herald* and *York Herald*, issue 14017, 9 May 1896, 10.

14 *Cycling*, issue 400, 17 September 1898, 197.

15 *Punch*, 16 October 1897, 172.

16 *Hearth and Home*, issue 288, 19 November 1896, 50.

17 *Cycling*, issue 286, 11 July 1896, 462.

18 Ibid.

19 Ibid.

20 *Cycling*, issue 184, 28 July 1894, 22.

21 George Herschell, *Cycling as a Cause of Heart Disease* (London: Bailliere, Tindall & Cox, 1896). Quoted in *Cycling: The Craze of the Hour*, 50.

22 *Cycling*, issue 382, 14 May 1898, 414.

23 *Cyclists' Touring Club Gazette*, volume 15, issue 8, March 1896, 95.

24 Ibid.

25 See Sheila Hanlon, 'Bicycle Face: A Guide to Victorian Cycling Diseases', 4 January 2016. Retrieved from http://www.sheilahanlon.com/?p=1990

26 *Aberdeen Weekly Journal*, issue 7856, 3 August 1898, 8.

27 *Cycling*, issue 397, 27 August 1898, 132; *Cycling*, issue 323, 27 March 1897, 230.

28 *Cycling*, issue 15, 2 May 1891, 247.

29 *Cycling*, issue 208, 12 January 1895, 494.

30 *Tottenham Cyclist*, volume 1, issue 6, August 1897, 1-2.

31 *Cycling*, issue 346, 4 September 1897, 162.

32 *Cycling*, issue 347, 11 September 1897, 185.

33   *Punch*, 22 August 1896, 96.

34   *Leeds Evening Express*, 10 July 1897, 3; *Leeds Evening Express*, 16 July 1898, 3.

35   *Leeds Evening Express*, 8 June 1897, 1.

36   *Cycling*, issue 181, 7 July 1894, 391.

37   *Cycling*, issue 346, 4 September 1897, 162.

38   *Cycling Magazine*, October 1896. Quoted in Jeanne Mackenzie, *Cycling*, 33.

39   Michael Bunce, *The Countryside Ideal: Anglo-American Images of Landscape* (London: Routledge, 1994), 114.

40   *Clarion Cyclists Journal*, volume 1, issue 5, March 1897, 5.

41   *Manchester Guardian*, 21 August 1895, 8.

42   John Clive, 'The Use of the Past in Victorian England', *Salmagundi* 68/69, Fall 1985–Winter 1986, 48-52.

43   Mark Girouard, 'A Return to Camelot', *The Wilson Quarterly* 5, Autumn 1981, 182 (178-89). It is worth noting that Bernard Shaw's description of how he 'hit the dust like the Templar before the lance of Ivanhoe' was a direct reference to the climax of one of Scott's novels, while the club captain who dreamt of the 'happy times of yore' when cyclists safely travelled along the 'highways and bye-ways of merrie England' clearly demonstrates Scott-like imagery.

44   Quoted in *Edwardian Excursions, from the Diaries of A.C. Benson 1898-1904*, ed. David Newsome (London: John Murray, 1981), 118.

45   Ibid., 30.

46   See for instance 'Cycling Hobbies', *Hearth and Home*, issue 260, 7 May 1896, 997.

47   Gerry Moore, 'Cycling and Photography', *The Boneshaker: The Journal of the Veteran-Cycle Club*, volume 15, issue 147, Summer 1998, 4-7; Bob White, 'Cycling and Photography', *The Boneshaker: The Journal of the Veteran-Cycle Club*, volume 16, issue 150, Summer 1999, 21-9.

48   *Cycling*, issue 63, 2 April 1892, 168.

49   *Cyclorn*, volume 3, issue 2, May 1896, 3.

50   *Cycling*, issue 237, 3 August 1895, 34.

51   E.H. Lacon Watson, 'Bicycle Tours – And a Moral', *Westminster Review* 142, July 1894, 171.

52   *Cyclists' Touring Club Gazette*, volume 15, issue 8, March 1896, 80.

53   R.J. Mecredy, 'Cycling', *Fortnightly Review* 50, July 1891, 80.

54   *Cycling*, issue 100, 17 December 1892, 390.

55   *Cycling*, issue 375, 26 March 1898, 256.

56   *Stanley Gazette*, volume 8, issue 91, August 1899, 61; *The Stanley Gazette*, volume 8, issue 92, September 1899, 64-8.

57   *Stanley Gazette*, volume 8, issue 92, September 1899, 66.

58   *Cycling*, issue 144, 21 October 1893, 222.

59   *Cycling*, issue 211, 2 February 1895, 37.

60   *Cycling*, issue 21, 13 June 1891, 334.

61   Ibid.

62   George Lacy Hillier, *The Potterers Club*, 9.

63   *Lady Cyclists' Association News*, issue 23, November 1898, 5. Hull History Centre, U DX113/13.

64   Peter Zheutlin, *Around the World on Two Wheels: Annie Londonderry's Extraordinary Ride* (New York: Citadel Press Books, 2007). See also Peter Zheutlin, 'Chasing Annie', *Bicycling*, May 2005, 64-9. Retrieved from http://www.annielondonderry.com/images/BI05ANNIE.pdf. Debbie Foulkes, 'Annie Kopchovsky Londonderry (1870?-1947) Rode a Bicycle Around the World. Forgotten Newspapers'. Retrieved from https://forgottennewsmakers.com/2010/03/09/annie-kopchovsky-londonderry-1870-1947-rode-around-the-world-on-a-bicycle/

65   N.G. Bacon, 'A Pioneer Ride in a Cycling Dress', *The Review of Reviews*, October 1894, 406. See also 'Through the Air on Wheels: Interview with Miss Bacon', *The Woman's Signal*, issue 37, 13 September 1894, 168.

66   *Girl's Own Paper*, issue 886, 19 December 1896, 182.

67   Ibid.

68   Ibid.

69 *Freeman's Journal and Daily Commercial Advertiser*, 30 August 1899, 4.

70 *Manchester Guardian*, 21 August 1895, 8.

Conclusion

1 A.J.P. Taylor, *Essays in English History* (London: Penguin Books, 1991), 9.

2 Quoted in *Cycling*, issue 429, 8 April 1899, 260.

3 See Greg Buzwell, 'Daughters of Decadence: The New Woman in the Victorian fin de siècle', 15 May 2017. Retrieved from https://www.bl.uk/romantics-and-victorians/articles/daughters-of-decadence-the-new-woman-in-the-victorian-fin-de-siecle

4 *Cyclorn*, volume 3, issue 6, September 1896, 5.

5 Quoted in John Lowerson, *Sport and the English Middle Classes* (Manchester: Manchester University Press, 1993), 120.

6 John Lowerson, 'Sport and the Victorian Sunday: The Beginnings of Middle-Class Apostasy', in *A Sport-Loving Society*, ed. J.A. Mangan (London: Routledge, 2006), 187.

7 *Lady Cyclist*, 22 August 1896. Quoted in Andrew Ritchie, *King of the Road*, 162.

8 Andy Cope, 'Can We Put a Figure on the Value of Cycling to Society?', 17 March 2016. Retrieved from https://www.sustrans.org.uk/blog/can-we-put-figure-value-cycling-society

9 Raymond Williams, *Culture and Society 1780-1950* (Harmondsworth: Penguin Books, 1961), 323.

Index

Index